DARWINIAN
RACISM

DARWINIAN RACISM

HOW DARWINISM INFLUENCED HITLER, NAZISM, AND WHITE NATIONALISM

RICHARD WEIKART

SEATTLE   DISCOVERY INSTITUTE PRESS   2022

Description

To hear some tell it, Adolf Hitler was a Christian creationist who rejected Darwinian evolution. Award-winning historian Richard Weikart shows otherwise. According to Weikart, Darwinian evolution crucially influenced Hitler and the Nazis, and the Nazis zealously propagated evolutionary theory during the Third Reich. Inspired by arguments from both Darwin and early Darwinists, the Nazis viewed the "Nordic race" as superior to other races and set about advancing human evolution by ridding the world of "inferior" races and individuals. As Weikart also shows, these ideas circulate today among white nationalists and neo-Nazis, who routinely use Darwinian theory in their propaganda to advance a racist agenda. *Darwinian Racism* is careful history. It is also a wake-up call.

Copyright Notice

Library Cataloging Data

Darwinian Racism: How Darwinism Influenced Hitler, Nazism, and White Nationalism by Richard Weikart

Cover design by Brian Gage

188 pages, 6 x 9x 0.4 inches and 0.6 lbs, 229 x 152 x 10 mm & 0.26 kg

Library of Congress Control Number: 2021951765

ISBN: 978-1-63712-009-5 (paperback), 978-1-63712-011-8 (Kindle), 978-1-63712-010-1 (EPUB)

BISAC: SCI034000 SCIENCE/History

BISAC: HIS014000 HISTORY/Europe / Germany

BISAC: SOC070000 SOCIAL SCIENCE/Race & Ethnic Relations

BISAC: POL042030 POLITICAL SCIENCE/Political Ideologies/Fascism & Totalitarianism

Publisher Information

Discovery Institute Press, 208 Columbia Street, Seattle, WA 98104

Internet: discoveryinstitutepress.com

Published in the United States of America on acid-free paper.

First edition, first printing, February 2022

ADVANCE PRAISE

With meticulous historical scholarship, Richard Weikart builds a convincing case that the Nazi worldview was influenced by Darwin's theory of evolution. Scrutinizing original documents from Nazi propaganda to school curricula to scientific journals, he amasses evidence that Darwinism played a prominent role in Nazi racial ideology. When Hitler massacred millions of people, in his twisted logic he was lending a helping hand to the evolutionary struggle for existence. In an intriguing final section, Weikart surveys literature from today's neo-Nazis, white nationalists, and alt-right proponents to show that these groups likewise invoke Darwinian theory to claim that their racism is scientific. As always, Weikart is thoughtful, thorough, and convincing.
   —**Nancy Pearcey**, professor and scholar in residence at Houston Baptist University, author of *Total Truth and Love Thy Body*, co-author of *The Soul of Science*.

This is an excellent book which is long overdue. In it Professor Weikart documents the close relationship between Nazism and social Darwinism. In doing so he uncovers important roots of contemporary racism while documenting the growth of "scientific racism," which for almost one hundred years dominated Western education, policy making, and society. Drawing on well-documented evidence he shows how Darwinism fueled the eugenics movement, racist attitudes, and a disregard for human dignity. This book is a must-read for anyone interested in today's world and the history of modern oppression as well as the terrifying history of the twentieth century and a driving force behind Nazism.
   —**Irving Hexham**, professor of religion, University of Calgary, author of *Understanding World Religions*

Richard Weikart brilliantly exposes some of the bad ideas that led to the Holocaust. He convincingly answers his critics who deny the deep influence of Darwin on Hitler. He also unveils the Darwinian influence on today's alt-right racism. Weikart compels us to learn from the dark past to avoid repeating it.
—**Michael Newton Keas,** lecturer in history and philosophy of science, Biola University; author of *Unbelievable: 7 Myths about the History and Future of Science and Religion*

Richard Weikart displays extensive knowledge of the relevant sources and the scholarly debates on Darwinism in Nazi Germany. He convincingly argues that Darwinian evolutionary theory was a core component of the Nazi worldview, and one that directly influenced Nazi policy. It's sure to become the standard work on the subject.
—**William Skiles,** assistant professor of history, Regent University

In this well-researched and powerfully written book, Richard Weikart shows how Hitler and his National Socialists operationalized nature red in tooth and claw. Here was social Darwinism on steroids. The historiography of this era will be significantly shaped by this persuasive account. A must-read!
—**Michael A. Flannery,** professor emeritus of UAB Libraries, University of Alabama at Birmingham, author of *Nature's Prophet: Alfred Russel Wallace's Evolution from Natural Selection to Natural Theology*

Acknowledgments

A S WITH ANY SCHOLARLY WORK, I AM DEEPLY INDEBTED TO MANY other scholars who have studied these issues before me. Their names are found in the citations throughout the book. I also want to thank those who read the manuscript before publication and provided me with encouraging comments or suggestions for revisions: Mike Flannery, Irving Hexham, Mike Keas, Nancy Pearcey, William Skiles, John West, and Jonathan Witt.

I would also like to thank the libraries and archives that facilitated my research. First and foremost was the Inter-Library Loan Department of my own university, California State University, Stanislaus. In addition, thanks to the Hoover Institution archive and library at Stanford University; the Akademie der Künste archive in Berlin; the Ernst-Haeckel-Haus archive in Jena; the Staatsbibliothek Preussischer Kulturbesitz archive in Berlin; the University of California, Berkeley library; the University of California's Northern Regional Library Facility; and the California State Library. Thanks also to Wilfried Ploetz, who allowed me to do research in the Alfred Ploetz papers.

Some of the material in this book I already published in a peer-reviewed scholarly journal: "The Role of Darwinism in Nazi Racial Thought," *German Studies Review* 36 (2013): 537–556.

Thanks also to Discovery Institute's Center for Science and Culture for supporting my work throughout the years. Jonathan Witt did an amazing job editing my manuscript and preparing it for publication.

Finally, thanks to God, my wife, and my children for blessing me in ways too many to count.

CONTENTS

INTRODUCTION

IT WAS THE SPRING OF 1999, A DENVER SUBURB. THE DAY, APRIL 20—Adolf Hitler's birthday. An eighteen-year-old white nationalist, Eric Harris, donned a shirt emblazoned with "Natural Selection" before heading off to high school. For weeks he had been preparing a special event in honor of the Führer. Together with a co-conspirator, Dylan Klebold, he planted a bomb in the Columbine High School cafeteria. Harris planned to shoot his fellow students as they fled the explosion. When the bomb failed to detonate, he and Klebold entered the school and opened fire, killing thirteen and wounding twenty-four before turning their guns on themselves.

Why was Harris—as are many white nationalists today—so eager to honor both Hitler and Darwin? Why did he think Darwin's theory of natural selection provided fodder for his white nationalist ideology? What were the connections?

If we delve into the ideology of Nazis, neo-Nazis, and white nationalists, we find that Darwinism—the view that species have evolved over eons of time through the process of natural selection—plays a fundamental role, shaping their views about race and society. Both Hitler and Harris, together with other Nazis and white nationalists, believed that Darwinian theory contributed several key ideas to their racist ideology. Most importantly, they thought Darwinism implied that different races had evolved to different levels, with some races superior and others inferior. Further, they believed that these allegedly unequal races were locked in an inescapable struggle for existence, in a competition to the death. Nazis and white nationalists consider it their mission to advance their own race in this universal racial struggle, even to the point of perpetrating violence against those deemed their racial enemies. In the Darwinian struggle for existence, someone has to die, after all.

Many Darwinists will protest that evolutionary theory does not necessarily lead to the conclusions that Nazis and white nationalists have derived from it. Most Darwinists today are not racists and do not go around staging mass murders. They represent a wide variety of political and social philosophies, including egalitarianism and democracy. Many of them uphold human rights. But we can grant this obvious point and still pursue a historical inquiry that is increasingly, and alarmingly, relevant: Did Nazis believe that Darwinism informed their worldview, and do present-day white nationalists believe this? Were they influenced by Darwinian theory, and, if so, how?

Those wanting to distance Darwinism from the Nazis need to stop ignoring the fact that the racial inegalitarianism of the Nazis in the early to mid-twentieth century was not all that distant from the racist attitudes and theories of many leading Darwinian biologists, anthropologists, and physicians. Darwin himself was racist and exulted in the European extermination of the "lower races," which he integrated into his theory of human evolution. Many other scientists likewise promoted racism on the basis of their understanding of evolutionary theory. If the Nazi perspective was a misinterpretation of Darwinism, it was a misinterpretation fostered by the Darwinian biologists themselves, not by non-scientists or fringe publicists.

Indeed, long before the Nazis came on the scene, Darwinian biologists, anthropologists, and other scholars—including Darwin himself— were insisting that Darwinism provided intellectual support for racism and even racial extermination (and some also saw it as justification for militarism, economic competition, abortion, and euthanasia).[1] For instance, the leading Darwinian biologist in Germany, Ernst Haeckel, stated in 1904, "The distance between the thinking soul of the cultured human and the thoughtless animal soul of the wild natural human is extremely vast, greater than the distance between the latter and the soul of a dog."² Haeckel believed that Europeans had evolved to a higher level than other human races, and this view was quite common among scientists.

One of the most important features of Darwin's theory was his proposed mechanism for evolution: natural selection through the struggle for existence. Darwin argued that the population of any species—including humans—grew faster than the food supply, leading to competition for scarce resources in which the fit—those better adapted to their environments—survived and reproduced, while the unfit perished. This notion of natural selection would have a profound influence on conceptions of race relations, because Darwin thought that some races—such as black Africans, native Americans, and others—were intellectually inferior to Europeans, and that this explained why they were being exterminated by Europeans. Thus, many Europeans in the nineteenth century construed Darwin's theory as justification for annihilating other races. After all, these other races must be the "unfit," and in the Darwinian scheme of things, the unfit perish.

Many leading biologists, psychiatrists, and physicians in the late nineteenth and early twentieth centuries thought that Darwinism also provided support for eugenics policies. Eugenics was a movement that aimed at improving human heredity. Many eugenicists were disturbed by the way that modern societies were protecting their weak and sickly members, allowing them to survive and even reproduce. They feared that this contravening of natural selection would lead to biological degeneration, rather than upward evolution. To rectify matters, eugenicists proposed policies that would encourage the more prolific reproduction of those they considered biologically superior, while preventing the reproduction of those deemed inferior. By the early twentieth century, compulsory sterilization of people with disabilities was one of the most popular eugenics measures. In 1907 Indiana passed the first compulsory sterilization law, and many other states followed suit in the decades thereafter.

Was Nazism Influenced by Darwinism?

MANY HISTORIANS recognize that Hitler was a social Darwinist, and some even portray social Darwinism as a central, guiding element of Nazi ideology.[3] Thus it is strange that a small number of historians seem to think that Nazis did not believe in human evolution at all. George Mosse in *The Crisis of German Ideology: Intellectual Origins of the Third Reich* (1964) argued that acceptance of human evolution was incompatible with Nazi racial ideology, because of the Nazi stress on the perfection and immutability of the German race.[4] Similarly, and more recently, Peter Bowler and Michael Ruse have argued that the Nazis rejected human evolution, because they upheld a fixed racial type and racial inequality.[5] Nowhere is this more pronounced than in the work of Daniel Gasman, whose book purportedly demonstrates that Hitler built his ideology on the social Darwinist ideas of Ernst Haeckel, but then in the conclusion claims that the Nazis did not like the idea of human evolution.[6]

But how is it possible to embrace social Darwinism, while rejecting Darwinism and human evolution? Anne Harrington has suggested that the Nazis liked some elements of Darwinism, especially the struggle for existence, but not human evolution from primates.[7] Robert Richards supports this position, claiming that Nazi racial ideas and practices "were rarely connected with specific evolutionary conceptions of the transmutation of species and the animal origin of all human beings, even if the shibboleth 'struggle for existence' left vaporous trails through some of the biological literature of the Third Reich."[8]

This position seems plausible at first glance, especially since Houston Stewart Chamberlain, a forerunner of Nazi racial ideology, embraced this position. But the claim runs aground when we examine Nazi racial ideology in detail. In particular, the following lines of evidence demonstrate overwhelmingly that Nazi racial thinkers embraced human and racial evolution: 1) Hitler himself believed in human evolution. 2) The official Nazi school curriculum prominently featured biological evolution, including human evolution. 3) The Nazi Ministry of Education encouraged schools to purchase books teaching evolution. 4) Nazi racial anthropologists, including SS anthropologists, uniformly endorsed human evolution and integrated evolution into their racial ideology. 5) Nazi eugenicists argued that Darwinism was an integral part of eugenics ideology. 6) Nazi periodicals, including those on racial ideology, included discussions of evolutionary theory, and some even overtly combated creationism. 7) Nazi materials designed to inculcate the Nazi worldview among SS and military men vigorously promoted evolution as an integral part of the Nazi worldview.

Evolutionary theory shaped Nazi thinking in multiple ways. First, almost all Nazi racial theorists believed that humans had evolved from primates. Second, they provided evolutionary explanations for the historical development of different human races, including the Nordic or Aryan race. (These two terms were usually used synonymously, but most Nazi racial theorists preferred the term Nordic.) Specifically, they believed that the Nordic race had become superior because harsh climatic conditions in north-central Europe during the Ice Ages had sharpened the struggle for existence, causing the weak to perish and leaving only the most vigorous. Third, they believed that the differential evolutionary development of the races provided scientific evidence for racial inequality. Fourth, they held that the different and unequal human races were locked in an ineluctable struggle for existence. Fifth, they thought that the way for their own race to triumph in the struggle for existence was to reproduce more prolifically than competing races and to claim more living space in which to further increase their numbers.

These five points—rooted in the view that humans and human races evolved and are still evolving through the Darwinian mechanism of natural selection and the struggle for existence—profoundly impacted Nazi policy. They formed the backdrop for pursuing eugenics, killing the disabled, aggressively seeking more "living space," and exterminating members of races deemed inferior. These are hardly peripheral issues.

Of course, Nazi racial ideology was not derived exclusively from Darwinism or evolutionary biology. Some of my critics have misinterpreted my position, claiming that in my previous work I provided a monocausal argument that solely blames Darwinism for Nazism and the Holocaust. I find this objection baffling, because in the introduction to *From Darwin to Hitler* I explicitly reject a monocausal interpretation: "The multivalence of Darwinist and eugenics ideology, especially when applied to ethical, political, and social thought, together with the multiple roots of Nazi ideology, should make us suspicious of monocausal arguments about the origins of the Nazi worldview."

Later in that same paragraph I state, "I recognize the influence of political, social, economic, and other factors in the development of ideologies in general and of Nazism in particular—but these topics are outside the scope of this study." Just because I confine my analysis to one cause does not mean that I think it is the only cause—especially since I explicitly state that I do not hold that position. I also state, "Nor am I making the absurd claim that Darwinism of logical necessity leads (directly or indirectly) to Nazism."[9]

I recognize that Nazism drew on many sources. Indeed, in my later book, *Hitler's Religion*, I discuss many non-Darwinian influences on Hitler's ideology, such as Schopenhauer, Nietzsche, and Wagner.

We can also take note of the French racial theorist Arthur de Gobineau, who wrote before Darwin published *The Origin of Species* and contributed the racist idea that the Aryan race was superior to all other races. He also claimed that racial mixing produced deleterious effects, leading many racial thinkers, including the Nazis, to oppose interracial coupling. Hatred of the Jews had a long history pre-dating Darwin, and Darwin never evinced any anti-Semitic inclinations. Also, Mendelian genetics (i.e., the idea that a sexually reproducing organism passes on its hereditary traits in discrete units, which may be dominant or recessive, without the influence of the environment or the activity of the parents) played a role in debates over racial ideology within the Nazi regime—especially concerning policies opposed to interracial coupling ("miscegenation").[10]

However, in the decades preceding Hitler's rise to power, many German racial theorists had synthesized Gobineau, Mendel, and anti-Semitism with social Darwinism. Nazi racial theory generally embraced this synthesis. Racial thinkers, such as Ludwig Woltmann and Ludwig Schemann, had synthesized Gobineau and Darwin long before Hitler's rise to power.[11] The leading anthropologist Eugen Fischer and the geneticist Fritz Lenz, both very influential figures in racial science during the Nazi period, embraced both Gobineau and Darwinism (as well as Mendelism). The historian Hans-Walter Schmuhl, an expert on Nazi racial science and eugenics, has perceptively noted that despite some contradictions between Gobineau's racism and social Darwinism, "Nonetheless toward the end of the nineteenth century formulations of Gobineauism and social Darwinism blended into syncretistic racial theories."[12]

Some leading anti-Semitic thinkers in early twentieth-century Germany, such as Theodor Fritsch and Willibald Hentschel, incorporated Darwinism into anti-Semitic ideology.[13] Thus, many Nazi racial theorists interpreted the opposition between the Nordic and Jewish races as an episode in the Darwinian struggle for existence. Mendelian conceptions of heredity were incorporated into evolutionary biology in Germany, too, so debates over heredity were often debates over evolutionary theory. When Nazi racial theorists promoted Mendelism, they were not spurning evolution. They were rejecting Lamarckism (the idea that organisms can pass on to their offspring acquired characteristics) and were embracing what came to be known as the neo-Darwinian synthesis.[14]

Thus, I have held and continue to hold that while Hitler and Nazism drew on many divergent intellectual sources, Darwinism was an important, fundamental part of the Nazi worldview. This book provides further corroboration for that thesis.

In Christopher Hutton's book on *Race and the Third Reich*, he captures the evolutionary thrust of the Nazi regime in his conclusion, where he states, "All the key elements of this [i.e., the Nazi] world-view had been constructed and repeatedly reaffirmed by linguists, racial anthropologists, evolutionary scientists and geneticists. Ludwig Plate [an evolutionary biologist] observed that 'progress in evolution goes forward over millions of dead bodies' (Plate 1932: vii). For Nazism, survival in evolution required the genocide of the Jews."[15] Darwin's theory, as Plate explained, was a theory involving mass death for the unfit so that the fit could evolve to higher levels. By massacring multitudes of people he deemed inferior, Hitler was merely trying to lend evolution a helping hand. From his twisted perspective, he thought his atrocities would lead to evolutionary progress.

Indeed, Hitler and the Nazis thought it was their moral duty to promote evolutionary progress by ridding the world of those they deemed weak, sick, or members of allegedly inferior races. Thus, social Darwinism provided part of the ideological underpinning to Nazi policies of eugenics, euthanasia, military expansionism, and genocide.

However, a few scholars and even more non-scholarly skeptics still insist that Hitler and the Nazis did not believe in evolution at all. Though in my own earlier scholarship I analyzed the impact of Darwinism on Hitler's own worldview, up to now no one has examined the prevalence of social Darwinist thinking among German scientists, teachers, and Nazi publicists. Chapters 3-6 of the present book do just this, illuminating the way that social Darwinism impacted racial ideology in the Third Reich by examining the Nazi biology curriculum, Nazi propaganda, and the racial ideology of Nazi Germany's prominent biologists and anthropologists. Along the way I note various objections raised against my thesis and respond. Many of those responses to objections are clearly labeled by their section subheadings. One can also consult the index at the back of the book, specifically the entry "Objections." There the pages are listed where I respond to various objections. In one case an entire chapter is dedicated to rebutting an objection. Chapter 7 refutes the misleading claim that the Nazis rejected the Darwinian biologist Ernst Haeckel and banned his books.

In Chapter 8, we turn to the contemporary scene and the United States, exploring how social Darwinism has continued to influence neo-Nazis and white nationalists. Indeed, many contemporary white nationalists simply recycle the early twentieth-century arguments for scientific racism that prevailed in Nazi Germany. They insist that races have evolved different mental abilities and moral qualities, and they think races are locked in a Darwinian struggle for existence that will result in either survival or extinction. As we will see, the problem of Darwinian-inspired scientific racism remains alive, however much we may wish to ignore it. It is neither mirage nor specter.

1. The Racism of Darwin and Darwinism

I**N 1881, TOWARD THE END OF HIS LIFE, DARWIN WROTE TO A COL-**league that the "more civilised so-called Caucasian races have beaten the Turkish hollow in the struggle for existence. Looking to the world at no very distant date, what an endless number of the lower races will have been eliminated by the higher civilised races throughout the world."[1] This was not just some offhand comment unrelated to his science. It reflected important elements of his theory of human evolution. Indeed, he articulated this same principle in his scientific study of human evolution, *The Descent of Man* (1871), where he claimed, "At some future period, not very distant as measured by centuries, the civilised races of man will almost certainly exterminate and replace throughout the world the savage races."[2] Not only racism, but racial extermination was an integral feature of Darwin's theory from the start.

This is a position that has been articulated by many historians of science.[3] Two prominent historians specializing in the history of Darwinism, Adrian Desmond and James Moore, mince no words about the racism inherent in Darwin's theory. In their magisterial biography of Darwin, they state, "'Social Darwinism' is often taken to be something extraneous, an ugly concretion added to the pure Darwinian corpus after the event, tarnishing Darwin's image. But his notebooks make plain that competition, free trade, imperialism, racial extermination, and sexual inequality were written into the equation from the start—'Darwinism' was always intended to explain human society."[4]

It might come as a surprise to some that Desmond and Moore include "racial extermination" in this list, since in a later book, *Darwin's*

Sacred Cause: How a Hatred of Slavery Shaped Darwin's Views on Human Evolution, Desmond and Moore emphasize Darwin's humanitarianism and portray his loathing of slavery as a fundamental influence on his view of human evolution.[5] However, if one actually reads *Darwin's Sacred Cause*, one may be surprised to find that—despite their primary thesis—Desmond and Moore have not at all changed their position about Darwin embracing racism and even racial extermination. They state therein:

> By biologizing colonial eradication, Darwin was making 'racial' extinction an inevitable evolutionary consequence.... Races and species perishing was the norm of prehistory. The uncivilized races were following suite [sic], except that Darwin's mechanism here was modern-day massacre.... Imperialist expansion was becoming the very motor of human progress. It is interesting, given the family's emotional antislavery views, that Darwin's biologizing of genocide should appear to be so dispassionate.... Natural selection was now predicated on the weaker being extinguished. Individuals, races even, had to perish for progress to occur. Thus it was, that 'Wherever the European has trod, death seems to pursue the aboriginal'. Europeans were the agents of Evolution. Prichard's warning about aboriginal slaughter was intended to alert the nation, but Darwin was already naturalizing the cause and rationalizing the outcome.[6]

Thus, despite stressing Darwin's opposition to slavery, Desmond and Moore freely admit that he saw genocide—something most of us would consider an even graver evil than slavery—as a progressive force in human evolution. He was thereby justifying the imperialist wars against aboriginal peoples that Europe was conducting in his time. (By the way, Darwin was not unique in embracing both abolitionism and racism, as quite a few nineteenth-century abolitionists were also racists.)

Desmond and Moore reinforce this point later in the book by quoting from a letter Darwin wrote to Charles Kingsley: "It is very true what you say about the higher races of men, when high enough, will have spread & exterminated whole nations." Desmond and Moore then provide this explanation of Darwin's sentiments that he expressed in that letter: "While slavery demanded one's active participation, racial genocide was now normalized by natural selection and rationalized as *nature's* way of producing 'superior' races. Darwin had ended up calibrating human 'rank' no differently from the rest of his society."[7] Darwin's theory thus provided justification, not only for racism, but for racial struggle and even genocide.

Victorian Racism: Common but Not Ubiquitous

How HAD Darwin come to embrace these racist views? As many scholars have pointed out, Darwin's view that races are unequal is unremarkable. Such racist ideas were circulating widely throughout Europe, both in scientific and popular circles, long before Darwin came on the scene. Many Europeans and Americans used these ideas to justify race-based slavery in the Americas, as well as the European conquest of other lands, such as Australia, New Zealand, the Americas, and later Africa.

However, not all British men and women in the nineteenth century embraced racism. Some prominent British intellectuals, missionaries, and church leaders believed that black Africans, for instance, were equal to Europeans and only needed the proper education and upbringing to attain the technological sophistication of the Europeans. The famous British missionary and African explorer David Livingstone not only rejected the notion that black Africans were unequal to Europeans, but also devoted his life to showing them love and compassion. He dedicated his energies to fighting against the slave trade, and he even expressed support for the Africans when they fought against British colonial encroachments.[8] No wonder Livingstone was beloved by Africans and is still fondly remembered by black Africans.[9] One of the most prominent British intellectuals in the nineteenth century, John Stuart Mill, likewise rejected the idea of racial inequality.[10] Mill, like many of his contemporaries, embraced environmental determinism, so he believed that humans were shaped primarily by education and upbringing, not by their biology and heredity. Finally, Alfred Russel Wallace, the codiscoverer of natural selection, also rejected racism and opposed the idea that non-European races were somehow closer to non-human animals than their European counterparts.[11]

The Voyage of the Beagle

NONETHELESS, MOST of Darwin's contemporaries were racist, and these, one may assume, exerted an influence on Darwin. But by his own account, his view of other races was shaped by firsthand experience as he circumnavigated the globe between 1831 and 1836 on the HMS *Beagle*. Moreover, in his *Autobiography* Darwin said the voyage "has been by far the most important event in my life and has determined my whole career."[12] Darwin formulated his theory of evolution through natural selection in the years immediately following, as he reflected on his impressions from the voyage.

In his journal from the voyage, he consistently portrayed the aboriginal peoples of South America, Australia, New Zealand, and elsewhere as unintelligent and inferior to Europeans. Like many of his contemporaries, he regularly referred to them as "savages." In his conclusion he stated that the most astonishing experience of his trip was to observe "man in his lowest and most savage state." These men, he claimed, do not appear to have any human reason. Darwin even depicted the three Fuegians on board the *Beagle* as mentally inferior, even though they had lived in England, learned some English, and adapted to European ways. His contempt for the natives living in Tierra del Fuego was palpable: "Viewing such men, one can hardly make oneself believe that they are fellow-creatures, and inhabitants of the same world." Darwin clearly believed that Europeans were intellectually superior to natives in South America, Australia, and other parts of the world.[13]

During this voyage Darwin also learned a great deal about the conflict between the colonizing Europeans and the indigenous peoples. In South America eyewitnesses informed him about the Spanish settlers' battles with the natives there. He also visited New Zealand and Australia, where the Europeans were supplanting the aboriginal populations. While he was in Australia, he reflected on these developments, stating, "Wherever the European has trod, death seems to pursue the aboriginal.... The varieties of man seem to act on each other in the same way as different species of animals—the stronger always extirpating the weaker."[14] Thus, Darwin, even before he formulated his theory of natural selection through the struggle for existence, already believed that human races were unequal and were pitted against each other in a struggle to the death.

At times one detects in his writing a twinge of conscience about this slaughter, but on the whole, he seems to have approved of it. When he summed up his thoughts about the impact of Europeans on other lands, he ended up glorifying the British imperialist project. He argued that great improvement had come to the South Sea islanders through "the philanthropic spirit of the British nation." After mentioning the "civilizing" process in Australia (which included, of course, the annihilation of the natives), Darwin concluded, "It is impossible for an Englishman to behold these distant colonies, without a high pride and satisfaction. To hoist the British flag, seems to draw with it as a certain consequence, wealth, prosperity, and civilization."[15] Clearly, Darwin viewed the European extermination of native peoples as a progressive development, because he considered the Europeans superior bearers of civilization.

Racism Serves a Theory, A Theory Serves Racism

WHEN DARWIN began compiling evidence for biological evolution in his notebooks in the late 1830s, he included human evolution in his ruminations. He indicated that when human races confront each other, they fight and struggle with each other for supremacy. He wrote that differences in intelligence usually settle this conflict, though in the case of black Africans, their "organization" (presumably meaning their immunity to diseases that ravaged Europeans who moved to Africa) gave them an advantage in their homelands. His comments here imply that he thought not only that some race are more intelligent than others, but also that blacks were inferior in their mental abilities.[16] In addition to embracing the idea of intellectual inequality among human races, Darwin also wrote in his notebooks that he believed that human races differ in their moral sense or conscience.[17]

Darwin's racist and imperialist attitudes were conventional for his time, but his use of racism to defend his theory of human evolution buttressed those attitudes in the decades to follow by providing scientific justification for racism among many of Darwin's followers. Racism was not just an incidental part of Darwin's evolutionary theory. Rather Darwin considered racial inequality crucial evidence for his theory. In order to convince his contemporaries of his theory of evolution, he knew he needed to demonstrate the great variety within any given species, while minimizing the gap between different species. When applied to human evolution, this meant that Darwin had to stress human inequality on the one hand, and human proximity to apes on the other. Racism provided fodder for this argument, because Darwin placed the black Africans and Australian aborigines close to the apes in his racial hierarchy, while deeming the white Europeans far superior.

To be sure, when Darwin first published *On the Origin of Species* (1859), he mostly avoided the topic of human evolution. He understood that this was the most controversial part of his theory and that it would likely provoke resistance (as it did). As he explained twelve years later in the introduction to *The Descent of Man*, he had steered around the issue of human evolution "as I thought that I should thus only add to the prejudices against my views."[18] Only in the closing paragraphs of *Origin* had he briefly mentioned that his theory would likely have ramifications for human origins. Thus, when Darwin mentioned "races" in the full title of his 1859 book, *On the Origin of Species by Means of Natural Selection, or the Preservation of Favoured Races in the Struggle for Life*, he likely meant primarily varieties or sub-species of animals and plants, rather than human races. However, Darwin later clarified in *The Descent of Man* that he viewed human races as varieties or sub-species,[19] so everything he wrote in *Origin* did indeed apply to humanity. Darwin confirmed this in *The Descent of Man*, for one of its stated goals was to show that the evolutionary processes that Darwin had explained in *Origin* had brought about the origins of humans, too. *The Descent of Man*, in other words, argues quite explicitly for "the preservation of favoured" human "races in the struggle for life."

Darwin's conception of the struggle for life, or, as he more often called it, the struggle for existence, had highly problematic features when applied to humans. Darwin's signature theory of natural selection through the struggle for existence was based on Thomas Robert Malthus's population principle, which stated that humans (and other organisms) tend to reproduce faster than their food supply can increase. This implies that humans (and other species) are destined for mass death, since the food supply can never keep up with the ever-growing population. Darwin argued that because most organisms perish in their quest for limited resources, they are locked in an inescapable competition for those resources. This competition is most intense among members of the same species because they are competing for the same niche.

Despite the huge death toll resulting from the struggle for existence, Darwin considered it a positive force nonetheless, because it produced evolutionary progress. It weeded out the weak, sickly, and less capable— the "unfit"—while the "fit" survived and reproduced. In the last sentence of his chapter on "Struggle for Existence" in *The Origin of Species*, Darwin stated, "When we reflect on this struggle, we may console ourselves with the full belief, that the war of nature is not incessant, that no fear is felt, that death is generally prompt, and that the vigorous, the healthy, and the happy survive and multiply." Then, in the next-to-the-last sentence of the book, he stated, "Thus, from the war of nature, from famine and death, the most exalted object which we are capable of conceiving, namely, the production of the higher animals, directly follows."[20] When applied to humans, this would mean that humans are contending with their fellow humans for scarce resources in a competition-to-the-death. The fittest humans will survive and reproduce, while the less fit will die.

In *The Descent of Man* Darwin confirmed that he thought race played a central role in this struggle, so racism is not an incidental element of the book. Darwin explained from the outset the three main objectives of the work: 1) investigate whether humans are descended from some other animals; 2) explain the process of human evolution; and 3) describe "the value of the differences between the so-called races of man."[21] Of the seven chapters covering human evolution, one is entitled, "On the Races of Man," and racial themes also emerge in many of the other chapters.

Toward the beginning of the book's second chapter, "Comparison of the Mental Powers of Man and the Lower Animals," Darwin insisted that certain races were mentally inferior to others:

> Nor is the difference slight in moral disposition between a barbarian, such as the man described by the old navigator Byron, who dashed his child on the rocks for dropping a basket of sea-urchins, and a Howard or Clarkson; and in intellect, between a savage who does not use any abstract terms, and a Newton or Shakspeare [sic]. Differences of this kind between the highest men of the highest races and the lowest savages, are connected by the finest gradations. Therefore it is possible that they might pass and be developed into each other.[22]

Howard and Clarkson, incidentally, were leaders in the British abolitionist movement, and Darwin considered them the epitome of moral goodness. They were, of course, Europeans, as were Newton and Shakespeare, and clearly Darwin was identifying them as "the highest men of the highest races," in contrast to the "lowest savages." Thus, Darwin buttressed his theory of human evolution by asserting that Europeans were not only intellectually superior, but also higher on the scale of morality. This is highly ironic, of course, because these allegedly morally superior Europeans were at the time exterminating the supposedly morally inferior natives of the Americas, Australia, and elsewhere. Darwin apparently had no conscience about genocide, since he saw nothing amiss about allegedly morally superior people killing off those they deem inferior.

He considered the intellectual superiority of Europeans so self-evident that he wrote in a later chapter, "The variability or diversity of the mental faculties in men of the same race, not to mention the greater differences between the men of distinct races, is so notorious that not a word need here be said."[23] Despite its apparent obviousness (to him), however, later in the same chapter he did write more about it. He trotted out scientific evidence for intellectual disparities among races that he (and many other European scientists) considered compelling: the difference in their cranial capacities. According to the data cited by Darwin, the Europeans have the largest cranial capacities at 92.3 cubic inches, while Asians have 87.1 cubic inches, and Australians have only 81.9 cubic inches.[24] The lesson appeared unarguable: Europeans have greater intellectual abilities than do other races. Darwin used this same line of evidence to argue that women are intellectually inferior to men. (It should be noted that cranial capacity measurements cited above turned out to be inaccurate and misleading, and the relationship between cranial capacity and intelligence has been found to be neither straightforward nor well correlated.[25]) Later, when discussing the gap between present-day humans and simians, Darwin mentioned that the gap would only increase as the "savage races" were exterminated, because the black Africans or Australian aborigines were currently the closest races to the gorilla, which he considered the highest of the ape species.[26]

In a four-page section "On the Extinction of the Races of Man," Darwin explained that the primary cause of the extinction was a racial struggle for existence, which results in the decimation of weaker tribes and races. He claimed that the disappearance of ancient races was not the result of environmental factors or adverse circumstances. Rather, he averred, "Extinction follows chiefly from the competition of tribe with tribe, and race with race." Though disease may aid some people in these racial competitions, direct killing is also involved, because "when one of two adjoining tribes becomes more numerous and powerful than the other, the contest is soon settled by war, slaughter, cannibalism, slavery, and absorption." Darwin thought that in most cases the so-called civilized peoples were winning this bloody contest: "When civilised nations come into contact with barbarians the struggle is short, except where a deadly climate gives its aid to the native race."[27]

What shouldn't be overlooked here is that from Darwin's perspective, this pattern of natural selection by racial extermination was the path to human progress.

Scientific Racism in the Wake of Darwin

AMONG DARWIN'S followers in the nineteenth century, most agreed with his racist views, and nowhere was this more obvious than in Germany. Indeed, Ernst Haeckel, the leading German Darwinian biologist in the late nineteenth and early twentieth centuries, was even more racist than Darwin. While Darwin insisted that humans comprised a single species, Haeckel considered differences between races so acute that he assigned them to ten distinct species. If this was not enough to accentuate human inequality, Haeckel further suggested that these ten species be divided further into four separate genera.[28]

In *The Natural History of Creation*, published in 1868, three years before Darwin's *Descent of Man*, Haeckel included a diagram (Figure 1.1) containing heads of different human races and different simian species. His illustration, he explained, showed that "the differences between the lowest humans and the highest apes are smaller than the differences between the lowest and the highest humans."[29] Thus, Haeckel thought black Africans and Australian aborigines stood closer on the evolutionary scale to apes than to Europeans.

In his 1904 book, *The Wonders of Life*, Haeckel devoted an entire chapter to "The Value of Life." He argued therein that not all human lives have the same value. So how did he determine which had more or less value? Primarily by race. After articulating his dim view of the so-called primitive human races, whose cognitive abilities he compared with apes and dogs, he then stated, "The value of life of these lower wild

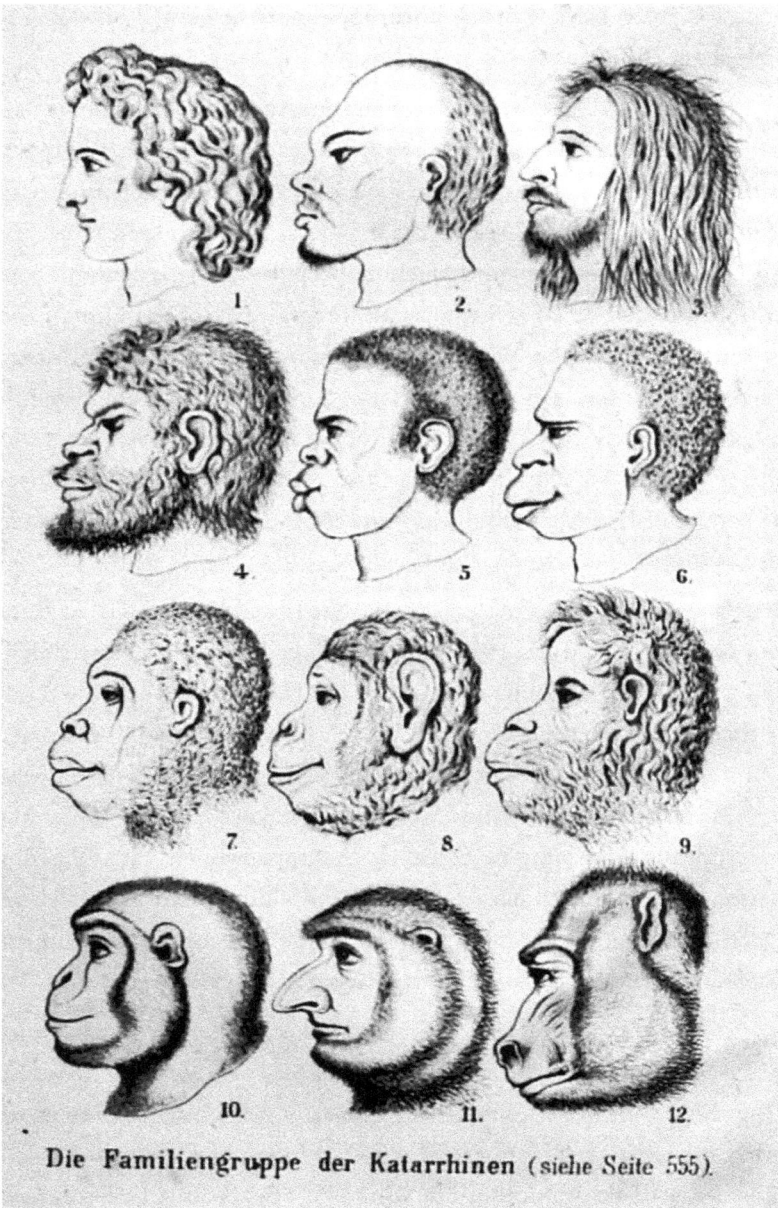

Die Familiengruppe der Katarrhinen (siehe Seite 555).

Figure 1.1. This frontispiece of German Darwinist Ernst Haeckel's 1868 book on evolutionary theory showed six human races and six simian species. The depiction of the "lowest" human in the diagram is suspiciously similar to the "highest" ape species.

peoples is equal to that of the anthropoid apes or stands only slightly above them."[30]

Haeckel, like Darwin, believed that the struggle for existence is most intense between members of the same species, so the primary competitor of a human is a fellow human. Haeckel explained that this struggle inevitably leads to the death of those who are less fit, even when it is waged in a non-violent manner, such as through peaceful economic competition. He saw racial struggle as an integral part of the human contest for supremacy. The Native Americans, black Africans, and Australian aborigines were examples, he claimed, of races that were losing the struggle for existence to the white Europeans.[31] "Even if these races were to propagate more abundantly than the white Europeans," he mused, "yet sooner or later they would succumb to the latter in the struggle for existence."[32]

It should not be surprising, then, that Haeckel was a fan of European colonialism. In his 1917 book, *Eternity*, he made it clear that he supported German colonization of Africa. There he also expressed horror that Germany's enemies in World War I were using non-European troops against them, calling this an "underhanded *betrayal of the white race*." These members of "wild" races simply did not have the same value as Europeans, according to Haeckel: "A single well-educated German warrior, though unfortunately they are now falling in droves, has a higher intellectual and moral value of life than hundreds of the raw primitive peoples, which England and France, Russia and Italy set against us."[33]

Another Darwinist who promoted the European colonial project was Friedrich Ratzel, who contributed to later Nazi ideology by formulating the idea of *Lebensraum* (living space), which Hitler used to justify his expansionist agenda. Ratzel was a Darwinian biologist who in 1869 published a book promoting Darwin's new theory. In this book, published two years before Darwin made similar arguments in *The Descent of Man*, Ratzel argued that the extermination of "primitive peoples" by Europeans was a powerful example of Darwinian natural selection in operation: "And can we any longer doubt the existence of natural selection," he commented, "when we read, how the last remnant of primitive peoples melt like snow in sunshine, as soon as they come into contact with the European, and how the European peoples for three centuries have populated entire large continents?"[34]

Ratzel, who later changed careers to become a geography professor, believed that the most important element in the struggle for existence was the "struggle for space." Based on this interpretation of Darwinism, Ratzel set forth a geographical theory that he hoped would become "nothing less than the foundation of a new theory of humanity." It focused on human migrations and the "struggle for space," or as he later termed it, the struggle for living space (*Lebensraum*).[35] Ratzel encouraged Europeans to wage wars of colonial conquest, in which they could "quickly and completely displace the inhabitants, for which North America, southern Brazil, Tasmania, and New Zealand provide the best examples." Though Ratzel did not promote biological racism in the same way that Darwin or Haeckel had, he did present the conflict between different peoples as part of a Darwinian "struggle for space."[36]

In the late nineteenth and early twentieth centuries scientific racism was widespread among biologists and anthropologists.[37] While Darwin, Haeckel, and their early followers used racism as a line of evidence to corroborate their theory of natural selection, later many Darwinists would turn the argument around. Instead of arguing that racism helped prove Darwinism, they would argue that Darwinism validated racial inequality. Thus, in its infancy Darwinism was influenced by racism, but soon it was being used to promote further racism by those who adopted it. While many scientists promoted social Darwinism and biological racism, a politician who took them seriously would pursue policies intended to advance his fellow Germans in the racial struggle for existence. Hitler hoped to foster evolutionary progress by promoting the so-called Aryan race, while exterminating those races he deemed inferior.

2. HITLER'S DARWINIAN WORLDVIEW

IN TWO PREVIOUS BOOKS I DETAILED THE INFLUENCE DARWINISM had on Hitler and the Nazis. The first is *From Darwin to Hitler: Evolutionary Ethics, Eugenics, and Racism in Germany*. The second is *Hitler's Ethic: The Nazi Pursuit of Evolutionary Progress*. While the works were strongly endorsed by several historians of the period, a few scholars challenged my historical case that Darwinism exerted considerable influence on the Nazi worldview. The two books were both peer-reviewed, scholarly works published by the respectable academic publisher Palgrave Macmillan. Most of the criticisms were raised by people who are not experts on Nazi ideology, and they often took isolated, idiosyncratic pieces of evidence and implied that they were the norm while ignoring much evidence that runs contrary to their claims. This chapter and the ones to follow will provide a wealth of evidence for my position, evidence that speaks against the objections raised. We begin with Hitler.

Traudl Junge, Hitler's last private secretary, testified that during the war, when Hitler was wont to regale his entourage with monologues, "occasionally interesting discussions arose about the church and human evolution." Hitler viewed Christianity negatively, she explained. "His religion was the laws of nature," she said. "He could submit to its violent dogma better than to the Christian doctrine of loving one's neighbor and one's enemy." According to Junge, during one of these discussions Hitler stated,

> Science is not yet clear about which branch humans originated from. We are obviously the highest stage of evolution of any mammal, that evolved from the reptile to the mammal, perhaps through the apes, up

to humans. We are a member of creation and children of nature, and the same laws are valid for us as for all living organisms. And in nature the law of struggle reigns from the beginning. Everything that is incapable of life and everything weak is eradicated. Only humans, and especially the church, have made it their goal to artificially preserve the weak, the unfit for life, and the inferior.[1]

These ideas, which Hitler also expressed many other times, were consistent with the pronouncements of many leading scientists and physicians in early twentieth-century Germany. They were not merely the musing of some fanatical fringe element in Germany (though some fringe movements did embrace these ideas, too).

Darwinian Influences on Hitler

WHERE DID Hitler imbibe this social Darwinist ideology? One possible source was Jörg Lanz von Liebenfels, a Viennese occultist who wrote pamphlets promoting social Darwinist racism, Aryan supremacy, and eugenics—ideas that would become integral to Hitler's worldview. Noticing the many striking parallels between Hitler's and Lanz von Liebenfels's ideology, the Austrian psychologist Wilfried Daim argued that Lanz von Liebenfels was *Der Mann, der Hitler die Ideen gab* (*The Man Who Gave Hitler His Ideas*).[2] However, Daim ignored two crucial factors: 1) There were significant differences between Hitler's and Lanz von Liebenfels's ideology, since Hitler never used the occult terminology so prevalent in Lanz von Liebenfels's writings; and 2) many scholars and writers in Vienna (and Munich and elsewhere in Germany) embraced social Darwinist racism, Aryan supremacy, and eugenics, so Hitler could have learned them from a wide variety of sources.

Indeed, it seems pretty clear that Hitler did not get all his ideas from any one individual. Hitler was eclectic. He loved Schopenhauer's and Nietzsche's philosophy, but he also admired Frederick the Great's Enlightenment rationalism. He embraced scientific racism and eugenics, views endorsed by many leading scientists and medical professors, but he also privately expressed support for the bizarre "World Ice Theory," which explained cosmology as an eternal struggle between fire and ice. Hitler loved listening to Wagner's operas, watching films, and reading the press. There were many sources shaping Hitler's worldview.[3]

One important influence helping shape, solidify, and confirm Hitler in his views about social Darwinism, racism, and eugenics was his friend Julius Friedrich Lehmann, a Munich publisher who specialized in scientific and medical books. Lehmann joined the fledgling Nazi Party in March 1920 and befriended Hitler, sending him copies of the books he published on racism and eugenics.[4] Perhaps even more importantly, in 1917 Lehmann began publishing *Deutschlands Erneuerung* (*Germany's Renewal*), which solicited articles from scientists, scholars, and publicists about racism, anti-Semitism, eugenics, German superiority, and similar themes. Lehmann's journal was so important to Hitler that in 1922 Hitler recommended that all Nazi Party members read it.

One of the leading Nazi ideologues, Alfred Rosenberg, published an article in the journal in 1922, and two years later Hitler did so as well, while he was in jail for his failed Beer Hall Putsch. Hitler's library contained an article by Fritz Lenz, who later became professor of eugenics at the University of Munich. In Lenz's essay, "Race as the Principle of Value: On a Renewal of Ethics," he argued that the "race ideal" is the highest ethical standard. He explained, "With every action, with every refraining from action, we have to ask ourselves the question: Does it benefit our race? And make our decision accordingly." This sounds remarkably like Hitler's later viewpoint, so it is not surprising that when Lenz republished this article as a pamphlet in 1933—the year Hitler came to power—Lenz boasted that it "contained all the main features of the National Socialist worldview."[5]

In addition to articles by Lenz, *Germany's Renewal* also published articles promoting eugenics by the University of Munich medical professor Max von Gruber and the plant geneticist Erwin Baur. In his 1922 article on "The Collapse of Cultured Peoples in the Light of Biology," Baur argued that the culprit for the decline and fall of civilizations was their neglect of biological laws. As he explained, natural selection—the law formulated by Darwin—eliminates the "inferior" elements of a population, keeping it robust; but cultured peoples reduce natural selection by helping the weak and sick, and "this elimination of natural selection is *the single* cause for the degeneration of cultured people." Baur also insisted that racial mixture was deleterious. One of the solutions he proposed to avoid biological degeneration was sterilization for those deemed biologically inferior.[6] In *Mein Kampf* and in his speeches, Hitler would later stake out positions in complete harmony with Baur's article.

Another author prominently featured in *Germany's Renewal* was Hans F. K. Günther, a leading Nordic racial theorist in the 1920s and 1930s (Nordic was a synonym for Aryan). Lehmann sponsored Günther's research and writing, publishing his books and giving copies of them to Hitler. In 1930, once the Nazis formed a coalition cabinet in the state of Thuringia, Hitler arranged for Günther to receive a professorship in anthropology. In one of his best-known books, *Rassenkunde des deutschen Volkes* (*Racial Science of the German People*), Günther honored Darwin as one of the most important men who "gave to human science the richest stimulus." He also explained that humans have evolved from more primitive human ancestors, such as *Pithecanthropus* and Neanderthals. He called on his fellow Germans to favor healthy Germans with Nordic ancestry by enacting policies such as banning marriage between Jews and Germans or promoting eugenics measures.[7]

Another book published by Lehmann that may have influenced Hitler—and that certainly showed strong affinities with Hitler's worldview—was *Human Heredity* (a two-volume work originally published as *Grundriss der menschlichen Erblichkeitslehre*) by Erwin Baur, the anthropologist Eugen Fischer, and Fritz Lenz. Baur, Fischer, and Lenz were all highly esteemed scientists in the 1920s. Baur was the editor of the journal *Zeitschrift für inductive Abstammungs- und Vererbungslehre* (*Journal for Inductive Evolutionary and Hereditary Theory*), and in 1928 he was given a prominent position as director of the Kaiser Wilhelm

Institute for Plant Breeding. Fischer was a professor of anthropology at the University of Freiburg, and then in 1927 he was promoted to one of the most prestigious posts for an anthropologist in Germany: director of the Kaiser Wilhelm Institute for Anthropology, Human Heredity, and Eugenics in Berlin. Lenz was appointed to a professorship in eugenics at the University of Munich in 1923, and then ten years later—after the Nazis took over the government—he joined Fischer at the Kaiser Wilhelm Institute.

The Baur-Fischer-Lenz work was well-received in scientific circles, becoming a standard text for courses on human heredity, not only in German universities, but in American and British universities as well (in its English translation, of course). The opening chapters, written by Fischer, emphasized racial inequality and the dangers of racial mixing.[8] Lenz, who wrote over three-quarters of the work, was even more emphatic about Nordic superiority. He claimed that people of the Nordic race were responsible for most cultural advances, creating the Persian, Greek, and Roman civilizations, and later contributing to most of the world's scientific and technological discoveries.[9] Lenz suggested introducing eugenics policies, including marriage restrictions, voluntary sterilization, legalization of abortion, and even euthanasia (defined by him as killing people with disabilities), in order to strengthen the Nordic race.[10]

Lenz also clarified the Darwinian basis for his eugenics ideology. The first chapter of the second volume is on "Human Selection." On the first page Lenz explains the Darwinian struggle for existence and how natural selection operates. He portrays eugenics as a way to help promote evolutionary progress, or at least to prevent biological degeneration.[11]

By examining the Baur-Fischer-Lenz text, it is apparent that social Darwinism, racism, and eugenics were mainstream science and not viewed in the 1920s and 1930s as "pseudo-scientific," a term used by some today to imply that it obviously has no place in the scientific enterprise. To be sure, these ideas were misguided and based on prejudices and false presuppositions. In this sense they are "pseudo-scientific" (though I prefer the term "misguided science"). However, most scientists in the 1920s and 1930s were blithely unaware of the huge problems with these views. The historian Heiner Fangerau has examined the reviews of the Baur-Fischer-Lenz work, and most of them—especially those by fellow German scientists—were laudatory.[12] Fangerau argues that the positive reception of the Baur-Fischer-Lenz text gave scientific imprimatur for Hitler's racial policies, including his anti-Semitism. Fangerau stated, "The criminal consequences of the National Socialist anti-Semitic terror find their origin at least in part in the Baur-Fischer-Lenz [book]."[13]

Despite the many parallels between this textbook and Hitler's ideology, it is difficult to know for sure if Hitler or other leading Nazi leaders actually read it, especially since it was pitched primarily for a scientific and medical audience. However, Lenz bragged in 1931 that Hitler had read the second edition while in Landsberg Prison from 1923 to 1924. This is not unlikely, because Lehmann regularly gave Hitler the books he published, and Hitler read a great deal in Landsberg, later asserting (surely hyperbolically) that he had received the equivalent of a university education there. Lenz boasted of the textbook, "Many passages from it are reflected in Hitler's turn of phrase [in *Mein Kampf*]. In any case he appropriated for himself the essential ideas of race hygiene and its importance with great mental responsiveness and energy."[14] Lenz thus recognized the similarities between his ideas and Hitler's, and at the time he seemed quite proud of his influence on Hitler.

The Centrality of the Struggle for Existence in Hitler's Worldview

FOR HITLER, the Darwinian struggle for existence between humans, especially the competition between different races, was central to his worldview and policies. In his writing and speeches, he constantly invoked the "struggle for existence" and "struggle for life," both terms Darwin used to describe biological competition. Hitler continually fulminated against the Marxist notion of "class struggle" and vowed to replace it with "racial struggle." In a 1927 speech he explained:

Politics is the striving and struggle of a people (*Volk*) for its daily bread and its existence in the world, just as the individual devotes its entire life to the struggle for existence, for its daily bread. And then comes a second matter, caring for future survival, caring for the child. It is the struggle for the moment and the struggle for posterity. And all thinking and all planning serve in the deepest sense this struggle for the preservation of life.[15]

Hitler thus emphasized that all his goals and policies were aimed at helping the German people succeed in the struggle for existence. He also explained this clearly in his two books, *Mein Kampf* and his *Second Book*, which was not given a title because it was never published in his lifetime. In both books he explained—just as Darwin had in *The Origin of Species*—that populations tend to increase faster than their food supplies, and this results in a competitive struggle between organisms, including humans, for scarce resources. Humans with superior traits would triumph in this struggle for existence, Hitler claimed, while the inferior would perish. Hitler believed that he and his fellow Germans needed to embrace the ways of nature, keeping their focus on winning the struggle for existence.

In *Mein Kampf* Hitler appealed to the laws of nature to defend his "folkish philosophy" and its emphasis on racial inequality and racial competition:

The folkish philosophy finds the importance of mankind in its basic racial elements. In the state it sees in principle only a means to an end and construes its end as the preservation of the racial existence of man. Thus, it by no means believes in an equality of the races, but along with their difference it recognizes their higher or lesser value and feels itself obligated, through this knowledge, to promote the victory of the better and stronger, and demand the subordination of the inferior and weaker in accordance with the eternal will that dominates this universe. Thus, in principle, it serves the basic aristocratic idea of Nature and believes in the validity of this law down to the last individual. It sees not only the different value of the races, but also the different value of individuals.[16]

The term that Hitler used here to describe nature—aristocratic—
was also used by the Darwinian biologist Ernst Haeckel when he bashed
socialism as an allegedly unscientific political position. Haeckel assured
his fellow Germans that the Darwinian process of natural selection was
based on human inequality and thus favored an inegalitarian society
(Haeckel was not supporting the traditional aristocracy, however, but a
meritocratic society).[17]

Earlier in *Mein Kampf* Hitler tackled the question of what to do
about Germany's growing population. He opposed birth control, because he thought it contravened Darwinian selection. He stated: "For
if conception as such is restricted and the number of births diminished,
then in the place of the natural struggle for existence, which only allows
the very strongest and healthiest to survive, comes the natural addiction
to 'save' at any cost even the weakest and sickliest; whereby the progeny is
set on a course, which must become ever more wretched, the longer this
mockery of nature and its will persists."[18] Instead of birth control, which
he thought would lead to biological degeneration, Hitler proposed expansionism as the best solution to Germany's population increase. Germany, he insisted, needed more living space (*Lebensraum*), just as the
Darwinian geographer Ratzel had suggested.[19]

Ideologically the most important chapter of *Mein Kampf* is "People
and Race" (*Volk und Rasse*), which was the only chapter to be reprinted
as a propaganda pamphlet for use in schools and Nazi organizations
during the Third Reich. In that chapter Hitler claimed that interbreeding of human races is counterproductive, because it would not promote
evolutionary progress. He stated, "The stronger must rule; it must not
unite with the weaker, thus sacrificing its own stature. Only the born
weakling can think this cruel, and that is why he is a weak and defective
man; for if this law did not hold, any conceivable evolution of organic living things would be unthinkable." Rather than intermingling with the
so-called inferior races, Hitler called for a competitive struggle between
races, because he thought this was in harmony with nature. He argued,

"Always struggle is a means to improve the health and stamina of the species, and thus a cause of its evolution. By any other process all development and evolution would cease, and the very reverse would take place."[20]

In his *Second Book* Hitler continued to stress the primacy of the struggle for existence in his worldview. The first chapter, "War and Peace in the Struggle for Life," opened by arguing that the twin instincts of self-preservation and reproduction bring about population growth that necessitates a struggle for existence among humans. The next two chapters, "Struggle, not the Economy, Secures Life," and "Race, Struggle, and Power," expanded on this theme and focused attention on the racial struggle. As in *Mein Kampf*, he emphasized the need for gaining land to feed the burgeoning German population, stating that "in the limitation of this living space (*Lebensraum*) lies the compulsion for the struggle for existence (*Lebenskampf*), and the struggle for existence (*Lebenskampf*), in turn, contains the precondition for evolution."[21] Hitler hoped his expansionist policies would promote evolutionary progress by increasing the numbers of "superior" people, while eliminating "inferior" human beings.

Hitler's View of Human Evolution

HITLER'S STATEMENTS about human evolution, natural selection, and the struggle for existence were by no means restricted to *Mein Kampf* and his *Second Book*. For example, in a speech to ten thousand new military officers on May 30, 1942, he told them the war they were fighting was inevitable, because they had no choice but to obey the laws of nature, including the struggle for existence. His speech was laced with Darwinian terminology, such as evolution, struggle, and selection. He opened the speech with remarks divulging his social Darwinist mindset:

> A deeply serious principle of a great military philosopher states, that *struggle* and thus war is *the father of all things*. Whoever casts even a glance at nature as it is, will find this principle confirmed as *valid for all organisms and for all happenings* not only *on this earth*, but even far

beyond it. The entire universe appears to be ruled only by this one idea, that *eternal selection* takes place, *in which the stronger in the end preserves its life* and the right to life, and the weaker falls.[22]

He then told his audience that the struggle over territory pits one people against another and leads to an "eternal selection, to the selection of *the best and hardest*. Thus we see in this struggle an element of the formation of every living thing and even of life itself." By eliminating the weaker and strengthening the stronger, this struggle, Hitler continued, produces "evolutionary progress" (*Vorwärtsentwicklung*).[23]

Nor was such talk by Hitler anything new. Fifteen years earlier, in a 1927 speech, he berated pacifists, informing them that the "law of the eternal struggle" made mincemeat of their ideals. He then told pacifists: "You are the product of this struggle. If your ancestors had not fought, today you would be an animal. They did not gain their rights through peaceful debates with wild animals, and later perhaps also with humans, through the comparative adjustment of relations by a pacifist court of arbitration, but rather the earth has been acquired on the basis of the right of the stronger."[24] This clearly indicates Hitler's conception that the struggle for existence produced humans from animal ancestors.

After coming to power, at his closing speech to the Nuremberg Party Rally in 1933, Hitler stated, "The gulf between the lowest creature which can still be styled man and our highest races is greater than that between the lowest type of man and the highest ape."[25] This sentence was taken almost verbatim from the Darwinian biologist Haeckel. As noted in Chapter 1, the German biologist tried to make the idea of human evolution from apes more plausible by maximizing the differences between human races while minimizing the gap between the "lowest" humans and the apes.

In a speech to construction workers a few years later, Hitler explained that his policies and goals were in harmony with the laws of nature, especially the evolutionary law of natural selection. He told them,

It is absolutely true that first of all the law of selection exists in the world, and nature has granted the stronger and healthier the right to life. And rightly so. Nature knows no weakling or coward, it knows no beggar, etc., but rather nature knows only those who stand firm on their soil, who sacrifice their life, and indeed sacrifice it dearly, and not those who give it away. That is an eternal law of nature. You see it if you gaze into the forest, you see it in every meadow, you see it in the struggle of individual organisms in the world, and you see it through-out the millennia of human history.[26]

Hitler thus justified his aggressive political, diplomatic, and military posture as natural, necessary, and beneficial to the German people.

As his secretary Junge explained (see the opening passage in this chapter), Hitler explicitly discussed human evolution during his mono-logues during World War II. (Some of these monologues were tran-scribed at the time, and historians generally consider the German ver-sions reliable while the English translation is considered unreliable, because it is based on different—probably fraudulent—transcripts.) On October 24, 1941, in a lengthy monologue on evolution, science, and religion, Hitler expressed his support for evolutionary theory, which—in his view—undermined Christianity. At the end of his long talk, he said, "There have been humans at the rank at least of a baboon in any case for 300,000 years at least. The ape is distinguished from the low-est human less than such a human is from a thinker like, for example, Schopenhauer."[27]

Several months later Hitler chatted with his entourage about why men shave their beards. He made the rather bizarre speculation that shaving is "nothing but the continuation of an evolution that has been proceeding for millions of years: Gradually humans lost their hair."[28] Hitler clearly believed that humans had descended from some kind of hairy ape or ape-like creature.

And in a June 1944 speech to military officers Hitler divulged his social Darwinist outlook in blunt detail. He opened his speech by remarking that war is an inevitable phenomenon. Then he continued by declaring,

> Nature teaches us with every look into its working, into its events, that the principle of selection dominates it, that the stronger remains victor and the weaker succumbs. It teaches us, that what often appears to someone as cruelty, because he himself is affected or because through his education he has turned away from the laws of nature, is in reality necessary, in order to bring about a higher evolution of living organisms.[29]

This statement is already extremely clear in endorsing the idea of biological evolution through natural selection. In this speech he also reiterated his view that human-like organisms had only existed for a few million years and humans for about 300,000 years.

He then spelled out the implications of this evolutionary worldview for ethics. He warned these officers against practicing humanitarian ethics, since this would bring about human extinction, because other species would destroy us. He stated, "War is thus the unalterable law of all life, the precondition for the natural selection of the strong and simultaneously the process of eliminating the weaker. What appears to people thereby as cruel, is from the standpoint of nature obviously wise." Nature does not care about any abstract human rights, but judges solely according to the right of the strong, he explained.[30] Hitler was thereby endorsing the view that whatever promotes evolutionary progress is morally good, and whatever leads to biological degeneration is morally bad.

In both the 1943 and 1944 editions of a Christmas book intended to buoy the spirits of the German populace (who were rather war-weary by that time with many fearing defeat), the frontispiece contained a quotation from Hitler: "All of nature is a powerful struggle between power and weakness, an eternal victory of the strong over the weak." (See Figure 2.1.) By this time many Germans were probably hoping for a Christmas with "Silent Night" or "Peace on earth, goodwill toward men." Instead, they found Hitler justifying his militarism by his incessant appeal to engage in the natural struggle for existence.

DIE GANZE NATUR IST EIN GEWALTIGES RINGEN
ZWISCHEN KRAFT UND SCHWÄCHE,
EIN EWIGER SIEG DES STARKEN ÜBER DEN SCHWACHEN.

Figure 2.1. Hitler's message in a book on Christmas: "All of nature is a powerful struggle between power and weakness, an eternal victory of the strong over the weak."

The abundant evidence summarized above, I would argue, decisively demonstrates that Hitler embraced Darwinism and that Darwinism played a central role in his racist ideology of Aryan supremacy and violent domination. But there are a handful of objections to such a view in sufficiently wide circulation as to merit a response. Let's briefly consider those now.

Objection: Hitler Was a Creationist

MANY WEBSITES of religious skeptics advance this claim, basing it on the fact that Hitler used the term "Creator" in *Mein Kampf* and elsewhere. However, there are several major problems with this position. First, Hitler often conflated Creator (and God) with nature. This is why most translators of *Mein Kampf* capitalize nature throughout the book. I provide abundant evidence in *Hitler's Religion* that Hitler was a pantheist, which means that he believed nature is the same as God. Second, in his monologues Hitler explicitly embraced Darwinian evolution and scoffed at creationist explanations. In private Hitler continually dismissed Christianity and its view of God and miracles as outmoded and unscientific. Hitler's views were clearly not compatible with creationism. Third, even if Hitler did believe in a Creator separate from nature (and I do not believe he did), this does not mean he could not also believe in Darwinism. There are many theistic evolutionists today who believe that God created something or other at some time—for instance, the cosmos 13.7 billion years ago. If Hitler was a theist (which I doubt), then he was a theistic evolutionist, and this does not undermine my position in the slightest.

Objection: Some Major Influences on Hitler Rejected Darwinism

WHILE IT is true that some of the figures who influenced Hitler rejected Darwinism, this does not prove anything about what Hitler himself believed about Darwinism. Hitler was eclectic in his thought, drawing on many different writers. Some of these rejected Darwin (or even preceded Darwin, such as Schopenhauer). However, many of the key influences on Hitler's ideology did embrace Darwinism, such as Theodor Fritsch and Ludwig Woltmann. Ultimately, the only way to know whether Hitler believed in Darwinism is to examine Hitler's own words, not the ideas of his predecessors, even if they may have influenced his ideology in some way or other. As we have seen, an examination of Hitler's own writings, speeches, and private monologues plainly, and at multiple points, reveals Darwinism's seismic influence on his worldview.

Objection: In 1942, Hitler Denied Belief in Human Evolution

THIS IS the strongest piece of evidence that Hitler rejected human evolution. "Where do we get the right to believe that humanity was not already from its earliest origins what it is today?" he reportedly stated during a private conversation in January 1942. "Looking at nature teaches us that in the realm of plants and animals transformations and further developments occur. But never within a genus has evolution made such a wide leap, which humans must have made, if they had been transformed from an ape-like condition to what they are now."[31]

But we should note that while he was expressing doubt that humans evolved from ape-like creatures, he also assumed that plants and other animals evolved. Thus, he still believed in biological evolution for most organisms and likewise in the struggle for life that he repeatedly emphasized as normative. His comments at this time were so vague that we cannot tell how he thought humans had originated if they had not evolved.

Also, this was the only time Hitler expressed doubt about human evolution. As cataloged above, he explicitly embraced human evolution many other times in his life, including in private conversations. One expression of doubt about human evolution should not be taken to represent Hitler's position when he often (both before and after) stated his belief in human evolution and integrated evolutionary ideas into many facets of his worldview.

Conclusion

NOT ONLY was Darwinism a formative influence on Hitler's ideology, but it was also a central tenet driving his political and diplomatic decisions. To his followers he promoted a racist version of social Darwinism, both publicly and privately. He believed the struggle for existence promoted biological progress by eliminating the allegedly inferior races and purging the Aryan race of its supposedly retrograde elements, such as people with disabilities. Hitler wanted to cooperate with nature by helping those he deemed superior to thrive and reproduce, so the human race could advance in evolutionary development. Ironically, while committing some of the worst atrocities the world has ever known, Hitler not only believed that his deeds were morally righteous, but he supposed that he was contributing to biological and even moral progress.

3. Evolutionary Theory in Nazi Schools

WHAT DID THE NAZIS TEACH ABOUT EVOLUTION IN THE GERMAN schools under their control? If we want to know what position the Nazi regime took on Darwinian evolution—or on any topic with academic content—then a good place to start would be to examine the school curriculum during the Third Reich. Hitler's regime was zealous to bring every aspect of German society and culture under the control of the Nazi leadership. They called this process *Gleichschaltung,* which is often translated as "coordination," but in my opinion is better translated as "synchronization." Everything in German culture was to be in lockstep with their worldview. They especially focused on shaping the minds of the younger generation, so they tried to integrate their ideology into the German schools as quickly as possible. On December 18, 1933, the Prussian Ministry of Education, under the leadership of Bernhard Rust, a loyal Nazi who in May 1934 also took over the national Ministry of Education, decreed that the aim of schools was to "serve the nation in the spirit of National Socialism."[1]

Because of the centrality of biological racism in Nazi ideology, the regime dictated that biological instruction should be given special attention. On September 13, 1933, the Nazi regime mandated the teaching of heredity and racial science, both of which were closely tied—both in the Nazi worldview and in the prevailing scientific consensus—to Darwinian evolution. Teaching heredity and racial science was crucial, the Nazi regime insisted, because "the knowledge of fundamental biological facts and their application to each individual and group is a condition sine qua non for the renewal of our people. No pupil, boy or girl, should be allowed to leave school for life without this fundamental knowledge."[2] Indeed, German schoolchildren in the 1930s would be confronted quite often with the Nazi perspective on biology, and, as we shall see, Darwinian evolution was a prominent part of the curriculum.

Darwinism in the Nazi's Official Biology Curriculum

BIOLOGY IN the Nazi school curriculum was impregnated with Darwinian evolution. In 1938 the Nazi-led Ministry of Education published an official handbook for curriculum in the schools and included a chapter on biology. This official handbook mandated the teaching of evolution, including the evolution of human races, which it said evolved through "selection and elimination." And it stipulated, "The student must accept as something self-evident this most essential and most important natural law of elimination [of the unfit] together with evolution and reproduction." In the fifth class, teachers were instructed to teach about the "emergence of the primitive human races (in connection with the evolution of animals)." In the eighth class, students were to be taught evolution even more extensively, including lessons on "Lamarckism and Darwinism and their worldview and political implications," as well as the "origin and evolution of humanity and its races," which included segments on "prehistoric humanity and its races" and "contemporary human races in view of evolutionary history."[3]

The Ministry of Education's 1938 biology curriculum reflected the biology curriculum developed by the National Socialist Teachers' League in 1936–37, which likewise heavily emphasized evolution, including the evolution of human races. The Teachers' League document, authored by H. Linder and R. Lotze, encouraged teachers to stress evolution, because "the individual organism is temporary, the life of the species to which it belongs, is lasting, but is also a member in the great evolution of life in the course of geological times. Humans are also included in this life." Thus evolution was supposed to support the Nazis' collectivist ideals. Indeed, as Linder and Lotze repeatedly stated, biology was central to the

Nazi worldview, and the biology curriculum, which included large doses of evolution, was intended to propagate Nazi ideology.

In the upper grades collectivist ideology could be fostered by teaching the following: "Upon depicting the biological processes and lawfulness in the life of individual organisms, it follows as the crown of the entire curriculum a thorough treatment of reproduction, heredity and evolution, formation [i.e., evolution] of races, racial science, racial care, and population policy. Thus biology instruction leads to the biological-racial foundations of the People's Community and of the state leadership." This biology curriculum called for teaching plant and animal evolution and then human evolution, which included instruction on the origin of human races.[4]

One of the leading authorities on biology pedagogy in the Third Reich was Paul Brohmer, whose book *Der Unterricht in der Lebenskunde* (*Instruction in the Life Sciences*, 1943) was part of a series devoted to "National Socialist Pedagogy in School Instruction." Brohmer, a professor at a teacher's college in Kiel, claimed that Nazi ideology, which his views represented, was based on the laws of biology. After congratulating Darwin on inaugurating a "new, more fruitful era of biology," he criticized Darwin for the individualism inherent in his theory, which reflected English liberalism. Brohmer stressed, however, that his criticism was not directed against evolution per se, which he fully accepted. He merely believed that evolution should stress holism and collectivism rather than individualism.[5]

Another instructor of biology teachers, Ferdinand Rossner, in a book approved by the Nazi Ministry of Education, pressed for more coverage of evolution in biology classes.[6] Rossner was an instructor at the School for Teachers in Hanover, and he also worked for the Nazi Racial Policy Office in Hanover. In 1939–1941 he edited the three-volume *Handbuch für den Biologieunterricht* (*Handbook for Biology Instruction*). The second entry in the handbook was a ten-page article on human evolution by the prominent evolutionary anthropologists Hans Weinert and Gerhard

Heberer. Weinert and Heberer provided evidence for human evolution and called it a fact beyond dispute. Heberer also wrote the article on Darwin, wherein he honored the English scientist as a genius. (For more on Weinert and Heberer, see Chapter 4). Other articles in the handbook taught evolutionary theory, too, including the ones on evolution (*Entwicklungslehre*), "Selection and Elimination," Haeckel, and even one on Wilhelm Bölsche, a popularizer of evolutionary theory.[7]

Evolution in Biology Texts during the Third Reich

IF ONE examines biology texts published in Germany in the late 1930s and early 1940s—thus after the Nazis had some time to revise the curriculum to correspond to their ideological agenda—one finds that these texts provided extensive discussion of evolution, including the evolution of human races. One officially endorsed textbook was Jakob Graf's *Biologie für Oberschule und Gymnasium* (*Biology for Secondary and College-Preparatory School*). Graf had joined the Nazi Party in 1932 and played a leading role in the biology section of the National Socialist Teachers' League.[8] In the 1942 edition of his text, the fourth volume has an entire chapter on "Evolution and Its Importance for Worldview." Therein Graf combated Lamarckism and promoted Darwinian evolution through natural selection. The reason that Graf and his fellow Nazis rejected Lamarck's evolutionary theory was that they embraced what scientists called "hard heredity," a view that became consensus among scientists by the mid-twentieth century. In this view, biological traits (and at that time they generally included intellectual and moral traits in this category, too) are shaped entirely through hereditary factors that are not influenced by the environment. Lamarckism, on the other hand, implied that the environment could influence heredity. (Because of their affinity for environmental determinism, many Marxists in the early twentieth century—e.g., Stalin—embraced Lamarckism, leading Nazi racial theorists to connect Lamarckism with Marxism).[9]

In his textbook Graf also claimed that knowing about human evolution is important, because it shows that humans are not special among organisms. He also argued that evolution substantiates human inequality. In the subsequent chapter on "Racial Science," Graf spent about fifteen pages discussing human evolution and insisted that humans and apes have common ancestors.[10] Another officially approved biology text from 1940, Erich Meyer and Karl Zimmermann's *Lebenskunde: Lehrbuch der Biologie für höhere Schulen* (*Life Science: Biology Textbook for Secondary Schools*), also discussed human evolution. The authors stated, "In this hard time [the Ice Age] humans already lived. In the conflict with nature he improved physically and intellectually more and more. It [the Ice Age] bred him ever upward. We find him first as a half-animal prehuman, then as a primitive human who lived in caves and knew how to use fire and to make stone tools and hunting weapons."[11]

This was typical fare in biology texts during the Third Reich, where human evolution and the evolution of races were standard topics. The notion that the Ice Age had made the Nordic races superior was a common view among Nordic racists and thus was a widely held position among Nazi racial theorists.

A 1942 biology textbook by Hermann Wiehle and Marie Harm with the official imprimatur of the Reich Ministry of Education gave extended attention to human evolution. Of the ten main chapters, two were on evolution generally and another one was devoted exclusively to human evolution. Excluding the summary, review section, and appendices, the chapter on human evolution comprised over 14 percent of the main part of the text. One of the recommended activities for classes was a zoo visit to view the primates: "Since in the curriculum we have covered evolution and the origin of humanity, during a visit to the zoo the primates will especially grip us."[12] As this text and the accompanying activity make clear, German schoolchildren during the Third Reich were encouraged to see primates as their evolutionary relatives.

Darwinian themes are also prominent in Sepp Burgstaller's picture book, *Hereditary Theory, Racial Science, and Population Policy: 400 Picture Sketches for Use in Schools* (1941). Burgstaller explained in the introduction that these sketches were designed to promote Nazi ideals in the classroom. The heading on one page of pictures proclaimed: "Nature Eliminates Everything Sick and Weak. All Life is Struggle. The Weak Perishes." (See Figure 3.1) Halfway down the page, under illustrations depicting the biological struggle for existence, is another heading: "Only the Healthy and Vigorous Survive in the Struggle for Existence. Natural Selection." The heading on the following page declared, "Humans Disturb Natural Selection and Help the Sick and Weak. Contraselection." One of the illustrations about the negative impacts of "contraselection" was a graph that showed an increasing percentage of "inferior" people and a declining percentage of "valuable" people if society continued to allow the sick and weak to reproduce.[13]

Figure 3.1. Sepp Burgstaller's 1941 picture book for German classrooms teaches biology with the headings "Nature Eliminates Everything Sick and Weak. All Life Is Struggle. The Weak Perishes," and "Only the Healthy and Vigorous Survive in the Struggle for Existence. Natural Selection."

The Nazi Ministry of Education also occasionally published lists of books that school libraries should consider purchasing. Among these were quite a few that taught Darwinian theory. In 1936 the Ministry of Education approved University of Berlin zoologist Richard Hesse's book, *Abstammungslehre und Darwinismus* (*Evolutionary Theory and Darwinism*), which was entirely devoted to proving evolution. The edition published that year contained a chapter entitled "Evolutionary Theory Is Valid Even for Humans."[14] Later the same year the Ministry of Education also approved *Rassenpflege und Schule* (*Racial Care and School*), wherein the medical professor Martin Staemmler not only expounded on the racial struggle for existence and the role of *Lebensraum* (living space) in that struggle, but also explicitly taught neo-Darwinian evolution of human races through mutation and natural selection.[15] (For more on Staemmler, see Chapter 5).

The National Socialist Teachers' League

THE NATIONAL Socialist Teachers' League and SS organizations also encouraged biology teachers in the Third Reich to emphasize evolutionary theory. *Der Biologe* (*The Biologist*) was an official journal of the National Socialist Teachers' League from 1935 to 1939, before being taken over in 1939 by the SS Ahnenerbe, an organization devoted to studying and promoting Nazi racial theory. The editors of this magazine were so zealous in advocating evolutionary theory that they published many articles attacking creationists, both before and after the SS took it over. In 1938 they published an article by Weinert on human evolution, too. The editorial board of the journal from 1939 on included some of the leading shapers of Nazi racial ideology, including Karl Astel (University of Jena), Walter Gross (leader of the National Socialist Racial Policy Office), SS-Gruppenführer (Major General) Günther Pancke (head of the SS Race and Settlement Main Office), anthropologist Otto Reche (University of Leipzig), psychiatrist Ernst Rüdin (Kaiser Wilhelm Institute for Psychiatry), and others.[16]

The new editor of *Der Biologe* in 1939, SS officer and biologist Walter Greite, explained that the journal was henceforth serving the newly founded organization, the Reich League for Biology, which was affiliated with the SS organization Ahnenerbe and thus came under the governance of Heinrich Himmler as the Reichsführer SS. In a chart explaining the areas of research undertaken by the Reich League for Biology, the first category listed was "phylogeny," which is the study of the evolutionary descent of organisms. Anthropology was included as a specialty under this category.[17] Thus evolution, including human evolution, was front and center in their research program. Most of the articles opposing anti-evolutionism in *Der Biologe* focused their criticisms on the Catholic periodical *Natur und Kultur*, though Ferdinand Rossner in a 1939 article also criticized a few teachers in Germany who rejected evolution. Rossner claimed that anti-evolutionism was an indirect assault on Nazi racial ideology. He then insisted that the official Nazi curriculum unmistakably required the teaching of evolution in the schools. As far as he was concerned, the case was closed: anti-evolutionism was contrary to Nazi ideology.[18]

Upon reading Rossner's article, Konrad Lorenz, a member of the Nazi Racial Policy Office who later won a Nobel Prize for his work on animal behavior, expressed shock. He could hardly believe "that in the educational system of National Socialist Germany there are men even today [1940] who actually reject evolutionary theory and common descent as such." Lorenz refused to argue about the merits of evolutionary theory, because he considered that question already settled. Rather, the purpose of his essay was to show that evolutionary theory did not sweep away ideals, but rather served as a basis for them. Lorenz vociferously rejected Catholic otherworldly values, but he claimed that evolution provided an even more elevated ideal: the higher evolution of humanity. He further argued that teaching evolution is the best antidote for the Marxist belief in human equality. In his experience, the most committed National Socialists were those who understood and embraced evolutionary theory. Lorenz argued that the Christian command to love your neighbor as yourself is an evolutionary imperative, too: "Since for us the race and Volk are everything and the individual person as good as nothing, this command is for us a completely obvious demand." Lorenz clearly believed that evolutionary theory reinforced Nazi racial doctrines, including racial inequality and racial solidarity (collectivism).[19]

Objection: The Nazis Banned Pro-Darwin Books

A COUNTER to the above evidence involves calling attention to a list of categories of banned books published in 1935 in *Die Bücherei*, a periodical for librarians. This list, compiled for libraries in the state of Saxony, included several categories of works to be banned, including: "Works of worldview or biological character whose content is the superficial scientific enlightenment of a primitive Darwinism and monism (Haeckel and those emulating him, as well as Ostwald)."[20] Some critics insist that this shows the Nazis rejected Darwinism. It does not.

First, this statement does not ban Darwinism in toto, but "primitive Darwinism and monism." Some Nazis rejected Haeckel's and Ostwald's monism (as shown in Chapter 7) not because it contained evolutionary theory but because Haeckel's Monist League, especially in the 1920s and early 1930s, had tilted toward socialism, pacifism, feminism, and other doctrines contrary to the Nazi worldview.[21] Another reason some Nazis opposed monism was they thought it was a materialistic philosophy, which suggests that "primitive Darwinism" may have meant "materialistic Darwinism." In any case, this banned book list was not condemning Darwinism per se, but only what it saw as misinterpretations of Darwinism, which is why it used the term "primitive Darwinism."

Second and more decisively, most historians of the Nazi period recognize that the Nazi system was polycratic; Nazis had many disagreements among themselves. For example, historian Guenther Lewy, in *Harmful and Undesirable: Book Censorship in Nazi Germany*, shows that Nazis often disagreed among themselves about what books should be censored.[22] Thus, a Saxon state library official placing something on a banned book list is not decisive; it is evidence, but must be weighed against the flood of contrary evidence demonstrating a deep and broad commitment to Darwinism among the Nazis.

Finally, *Die Bücherei*, the same periodical that published the banned book list, also published book reviews and recommended books that libraries should buy. In 1934 they published an article on books dealing with race and eugenics. They recommended that libraries acquire books expressly teaching biological evolution, such as Staemmler's *Racial Care in the [German] Ethnic State*, the famous Baur-Fischer-Lenz work on human genetics and eugenics, and Günther's *Racial Science of the German People*. (I discuss these three works in Chapters 5, 2, and 4 respectively.)[23] In 1935 *Die Bücherei* compiled a list of the most significant books of that year, and one of them, *Hereditary Science, Racial Care, and Population Policy* contained an entire chapter on the evolution of species and races. (See my discussion of this in Chapter 5.)[24]

In 1937 *Die Bücherei* even listed Staemmler's *Racial Care in the [German] Ethnic State* as one of the top one hundred books acquired by German libraries.[25] This journal also gave its thumbs up to quite a few other works that promoted evolutionary theory.[26] The journal *Die Bücherei* not only apparently did not hold that books should be banned simply because they contain Darwinism or evolutionary theory, but even recommended that libraries acquire books teaching Darwinism.

The notion that the Nazis banned Darwinism is based on a misinterpretation of, and unwarranted extrapolation from, a single entry of a banned book list composed by a minor Saxon state official, together with ignorance of the avalanche of data showing that the Nazis approved of Darwinism and considered it vital to their worldview.

Conclusion

NOT ONLY did the Nazi biology curriculum teach Darwinian evolution, but it promoted human evolution as an integral part of Nazi ideology. German children had Darwinian evolution presented to them in their classrooms as evidence for human inequality, especially racial inequality. Most German biologists and anthropologists thought that different races had evolved to different levels, and this reinforced Nazi racial dogmas. It is also clear from the attacks on anti-evolutionism in the official journal for biology teachers that anti-evolutionism was not very prevalent, and the small number of teachers who embraced it were pilloried by the official Nazi organs.

4. DARWINIAN SCIENTISTS OF THE THIRD REICH

DURING THE Third Reich many German scientists, including in particular most German anthropologists, employed Darwinian arguments for racial inequality and even Nordic racism. These included those who had risen to prominence before the Nazis came to power. The Nazis feted them on many occasions and insisted that Nazi racial ideology had science on its side. When the Nordic racial theorist Hans F. K. Günther delivered his inaugural address at the University of Jena in 1930, his audience included more than just academics. Hitler was present, too, eager to hear one of his favorite anthropologists talk about one of his favorite topics: race.

Indeed Hitler had been instrumental in getting Günther appointed to the professorship in social anthropology, because the Nazi leader Wilhelm Frick was the Minister of Education in a coalition government in Thuringia (the German state in which Jena was located). Günther was not an academic anthropologist, and because of that, his appointment was controversial. While some prominent scientists wrote strong recommendations for him, others considered him unqualified for such an academic position (not because of his ideas, but because he lacked scientific research experience). The faculty at the University of Jena voted against his appointment, but Frick rode roughshod over their concerns and appointed him anyway, because Hitler favored it.

After his inaugural lecture, Günther spoke briefly with Hitler, who then departed. However, Nazi leader Hermann Goering, who had not attended the lecture, appeared for the celebratory dinner afterwards.[1]

Three years later, after the Nazis came to power, the regime showered Günther with honors. Günther had already joined the Nazi Party, and in the earliest phases of the regime, Nazi leaders recruited him to participate in government committees tasked with framing racial and eugenics policies. In 1935 the Nazi Ministry of Education appointed him to a professorship at the University of Berlin, one of the most prestigious academic posts in Germany. That same year, at the Nuremberg Party Congress, Günther was the first scholar to receive the State Prize of the Movement for Scholarly Achievements, a Nazi honor founded by Hitler to recognize his favorite scholars. Hitler's declaration about Günther, read by Alfred Rosenberg at the congress, praised Günther by stating, "In his many writings and above all in his *Racial Science of the German People* he laid the intellectual foundations for the striving of our movement and for the legislation of the National Socialist Reich."[2] On his fiftieth birthday in 1941, Günther received the Golden Party Badge, a very high Nazi Party honor.[3]

If anyone exemplified Nazi racial ideology, it was Günther, whose writings in the 1920s had exerted a powerful influence on Hitler and many of his colleagues. Günther's Nordic racism drew from a variety of sources, including the nineteenth-century French racial thinker Arthur de Gobineau, a pre-Darwinian writer who postulated that the so-called Aryan race (or Nordic race, to use the term Günther preferred) was a superior race, but had declined in vitality because of mixing with allegedly inferior races. Like many other Nordic racists in the early twentieth century, Günther believed Darwinian theory provided a powerful explanation for the origin of races and racial inequality. When Günther listed his intellectual forebears, he praised Darwin, the famous Darwinian biologist August Weismann, and the eugenicists Francis Galton, Alfred Ploetz, and Wilhelm Schallmayer as important intellectual predecessors. He also included the prominent social Darwinist thinkers Georges Vacher de Lapouge and Ludwig Woltmann, both of whom synthesized Darwinism with Gobineau's racial thought, as crucial influences on his thought.[4] Günther espoused human evolution, and he believed the Nordic race had originated in northern Europe and had spread through conquest.[5] Günther proposed racial hygiene (eugenics) to improve the Nordic race.[6]

Shortly after Hitler came to power, Günther lectured at the University of Jena on the role of heredity and selection in the state. The main point of the lecture was to explain the need for eugenics policies. He claimed that Darwin was a crucial influence on the development of modern scientific conceptions of heredity and selection, in part by supplanting Lamarckism (though he admitted that Lamarckian mechanisms might play some role in evolution). The state, he argued, needed to found its policies on the firm Darwinian basis of selection, rather than the Lamarckian teaching of environmental influence. He stated, "The only way to our goal is the Darwinian way, i.e., selection and elimination: The hereditarily valuable having many children, and the hereditarily inferior having few or no children." Günther then applauded the social Darwinist writers Otto Ammon and Alexander Tille for calling for a "social aristocracy," which would emerge by selecting those considered to have the most valuable biological traits.[7]

Leading Anthropologists before the Nazi Era

UNLIKE GÜNTHER, who was appointed to his professorship by the Nazis, Otto Reche became a professor of anthropology at the University of Leipzig in 1927 before the Nazis came to power. When he was studying at the University of Jena, he took courses from Ernst Haeckel, the leading German Darwinian biologist.[8] Like Günther, he was a Nordic racist and a devotee of Woltmann, a physician whose racist ideology was developed on a social Darwinist foundation. As a student Reche zealously read Woltmann's journal *Politisch-anthropologische Revue* (*Political-Anthropological Review*). He identified Woltmann as a "bold forerunner of the völkisch and the racial ideology, thus of the worldview that is the foundation of National Socialism."[9] In 1936 Reche republished some of Woltmann's books to make Germans aware of the contribution Woltmann had made to racial thought. In the foreword to his new edition of Woltmann's *Politische Anthropologie* (*Political Anthropology*), he noted that "every page was influenced by the spirit of Darwin."[10] Reche obviously subscribed to Woltmann's evolutionary view of racial anthropology, and this is evident in his own writings.[11]

Reche's racial views were almost identical to Nazi racial ideology. He was a strong proponent of eugenics and helped found two eugenics organizations before the Nazis came to power: the Vienna Society for Racial Care (*Gesellschaft für Rassenpflege*) in 1925 and the Leipzig branch of the German Society for Racial Hygiene in 1932.[12] He rejected Lamarckian evolutionary ideas, claiming that they were connected to Marxist thought.[13] In 1936 Reche published a book that attempted to demonstrate that the Nordic race arose in central and western Europe during the last ice age. He argued that the selective pressure of this colder period had shaped the hereditary traits of the Nordic race, elevating it above other races.[14] Reche once wrote:

> What we call "world history" is fundamentally nothing other than *the history of the Indo-Germanic people and its achievements*, the powerful, elevating and simultaneously tragic *heroic song of the Nordic race* and its idealism: a song, which informs us, how the power of the race creates the seemingly impossible and its hand stretches almost to the stars, and how this power is only too rapidly extinguished, where the "law of the race" is forgotten, where Nordic people no longer protect the purity of their blood and mix with races of lesser cultural ability.[15]

Reche had close ties to the Nazi Party long before he actually joined in 1937, and lectured to Nazi Party organizations on racial anthropology. In 1933–34 (and probably afterwards) Reche was one of the main instructors at three-day training seminars offered by the State Academy for Physicians' Continuing Education in Dresden for physicians and teachers, which indoctrinated four thousand professionals in 1933 alone with Nazi views about race and eugenics. In these lectures Reche expressed considerable enthusiasm for the Nazi regime, especially its racial ideology, which he taught. The first of Reche's three lectures was devoted entirely to human and racial evolution.[16] Later he eagerly offered his expertise to influence racial policy in the occupied Eastern territories during World War II.[17]

As with Reche, Eugen Fischer was a leading anthropologist even before the Nazis came to power. He was appointed professor of anthropology at the University of Freiburg in 1918, before the Nazi Party even existed. The prominence of Fischer and Reche before the Nazi seizure of power illustrates that social Darwinist racism was entrenched in German academic anthropology before the Nazis came to power. Indeed, (as mentioned briefly in Chapter 2), Fischer's racial ideas were likely influential—either directly or indirectly—on Hitler and other Nazis. In a 1910 published lecture Fischer gave to the Natural Science Society in Freiburg, he outlined racial ideas that later became the mainstay of Nazi ideology. In that talk he insisted that Gobineau and later thinkers, such as Woltmann, were correct in their key idea that races have different mental abilities. Fischer stated that the Nordic race, because it was the primary creator of culture, was the highest race.[18]

In Fischer's 1913 book on racial crossing, based on his studies in southern Africa of people descended from European men and black African women, he warned of the supposed perils of miscegenation.[19] Hitler's views in *Mein Kampf* mirrored Fischer's (ideas also being promoted by other racial thinkers at the time), since Hitler not only stressed the inequality of races, but also maintained that the Aryans were the only culture-creating race and that racial mixing was dangerous.

From the beginning of his career to the end, Fischer was committed to the view that human races were formed through Darwinian processes. In an essay from 1912–13 on "Races and Racial Formation" he asserted that racial variations were the basis for human evolution, and that natural selection determined which races would survive and which would go extinct.[20] In 1927 Fischer was even more explicit about his views on evolution in his work *Rasse und Rassenentstehung beim Menschen* (*Race and Racial Origins of Humans*). Therein he claimed that all humans had evolved from the same primate lineage. He believed that human evolution was driven by natural selection through the struggle for existence. Not only did human anatomy change during the evolutionary process, but human instincts, drives, and habits evolved as well, he claimed.[21] Toward the close of the Nazi period, in 1943, he wrote a book review of the evolutionary anthropologist Hans Weinert's *Der geistige Aufstieg der Menschheit* (*The Intellectual Ascent of Humanity*). He praised Weinert not only as a leading scholar of fossil humans, but also for his investigation of the mental evolution of humans. In the review he lambasted anyone who was foolish enough to dismiss "the firmly established proofs of the ape ancestry of humans." He called the denial of evolution laughable and unscientific.[22]

As director of the prestigious Kaiser Wilhelm Institute for Anthropology, Human Heredity, and Eugenics, Fischer cooperated with the Nazi regime and received considerable support, funding, and acclamation from Nazi officials.[23] In 1937 the Nazi regime put Fischer's antimiscegenation ideas into practice, when they secretly (and illegally) sterilized several hundred so-called "Rhineland bastards" in order to keep them from passing on their "inferior" racial traits. These were children who had been fathered by black African colonial troops in the Rhineland during the French occupation after World War I. Eugen Fischer and other leading anthropologists cooperated with this sterilization program by helping to identify the target population.[24] Two years later, on Fischer's sixty-fifth birthday, Hitler bestowed on him the Goethe Medal for Art and Scholarship.[25]

Evolutionary Anthropologists Promoted by the Nazis

WEINERT WAS another anthropologist given preferential treatment by the Nazi regime. He served both as a professor of anthropology and an SS officer. Specializing in evolutionary anthropology, he worked under Fischer at the Kaiser Wilhelm Institute for Anthropology, Human Genetics, and Eugenics until 1935, when the Nazi regime appointed him professor of anthropology at the University of Kiel. He published many books and articles during the Nazi period discussing the evolution of humans from primates and the evolution of races. In 1936 the Nazi Ministry of Education recommended that schools purchase his book *Die Rassen der Menschheit* (*The Races of Humanity*).[26] In the opening pages of this book, Weinert explained the importance of evolution for anthropology: "Anthropology, however, is the history of all humanity, beginning with its origin from anthropoid ape ancestors and continuing to the dividing and re-mixing of all contemporary human races."[27]

Later in the book Weinert claimed that the Nordic race had evolved to a higher level than other races, especially the Australian aborigines, whom he considered the lowest race.[28] He also explicated the process of racial evolution:

> Humanity did indeed originate a single time; however, the beings who first earned the name "early human" or "ape-man" were not completely the same. And this unequal value grew with the dispersion of humankind. Many races perished in the course of human evolution, eradicated or eclipsed by other races that conformed better to the demands of the environment. Today the culture of Europe is determinative for the continuance of the great races; further evolution will bypass any who cannot or will not embrace this culture.[29]

In his earlier book, *Biologische Grundlagen für Rassenkunde und Rassenhygiene* (*Biological Foundations for Racial Science and Eugenics*, 1934), Weinert had dedicated an entire chapter to human evolution and another to the evolution of human races. After applauding the Nazi Party for introducing compulsory sterilization, Weinert stated, "Today any fear of not being allowed on the basis of national-political considerations to advocate evolutionary theory is completely unnecessary."[30] He then remarked that no serious scientists doubted the primate ancestry of humans. He provided a sketch of an evolutionary tree (Figure 4.1) that showed humans and modern apes coming from an earlier common ancestor.

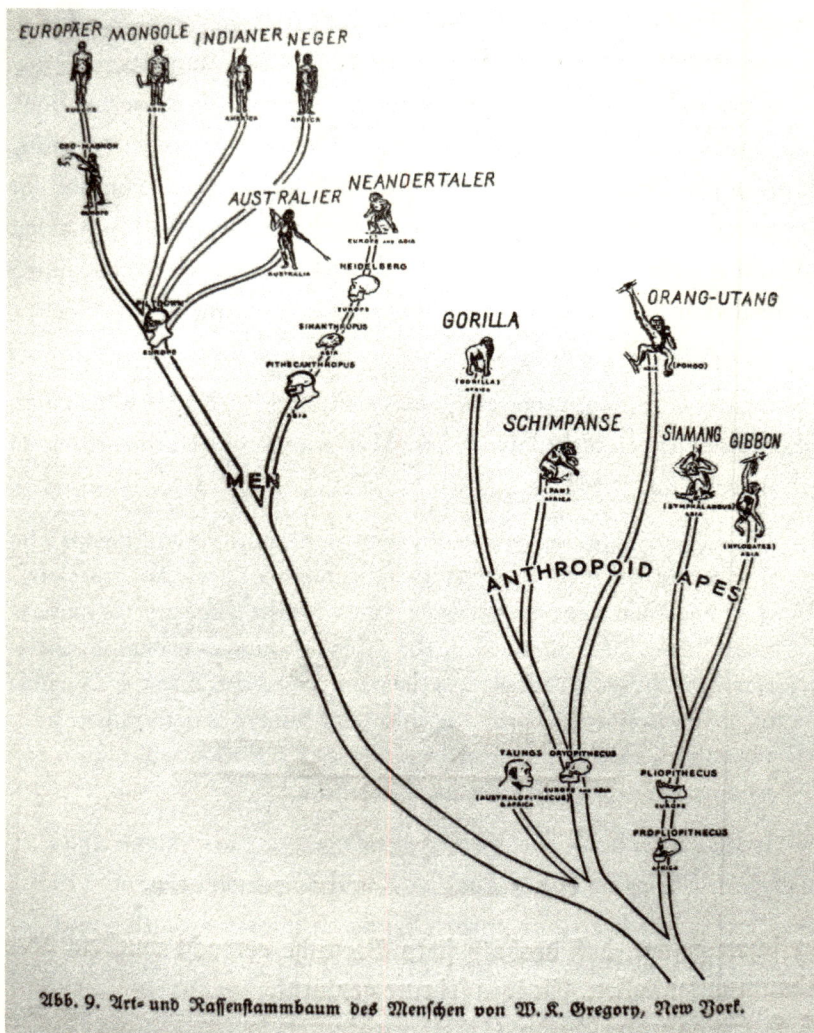

Figure 4.1. Phylogenetic tree of human races, in anthropologist Hans Weinert's *Biologische Grundlagen für Rassenkunde und Rassenhygiene* (1934).

In his chapter on the evolution of races he explained that despite common ancestry, "these races also have different *value*. The scientific theory of the common origin [of races] offers no foundation for a political thesis of the equal value of all humans!"[31]

In the 1941 edition of his book on the origin of human races, Weinert included a diagram that not only showed humans and apes as having
common evolutionary ancestry, but also placed humans under the category of apes. (Figure 4.2) He concluded this work by discussing racial
policy that would foster evolutionary progress. He warned against racial
mixing, and he hailed eugenics as the path to higher evolution.[32]

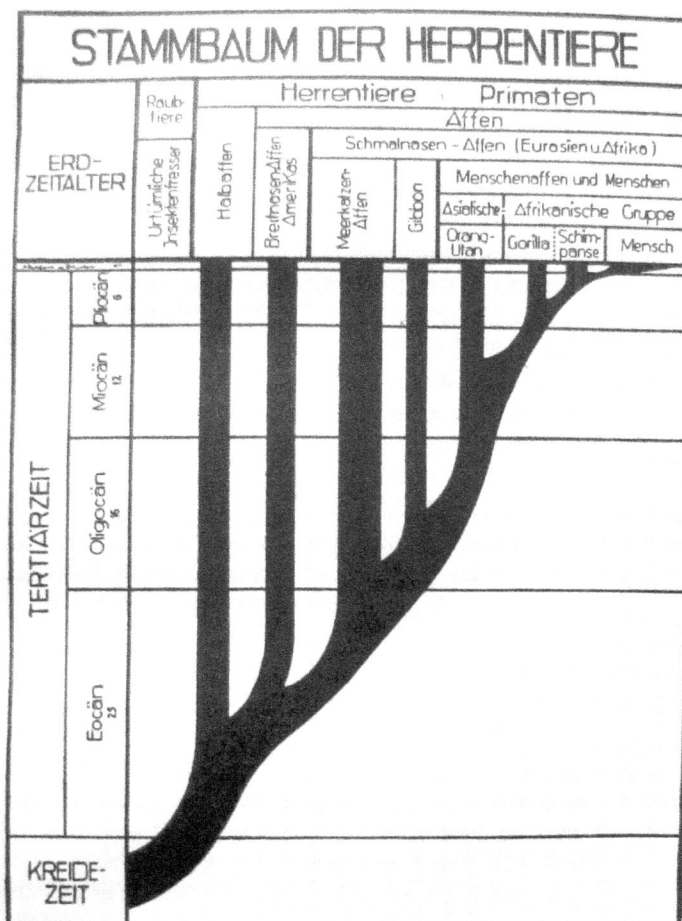

Figure 4.2. Evolutionary tree of primates that includes humans, in anthropologist Hans Weinert's *Entstehung der Menschenrassen* (1941).

Weinert's views on the evolution of human races were for the most part well-received by Nazi officials. The official periodical of the National Socialist Racial Policy Office listed Weinert's books, including *Die Rassen der Menschheit*, as valuable books on racial theory.[33] *Nationalsozialistische Monatshefte*, edited by Alfred Rosenberg, published a brief article in 1937 promoting one of Weinert's books on human evolution.[34] Ute Deichmann in *Biologists under Hitler* asserts that Walter Gross, the head of the Racial Policy Office, "considered Weinert the most competent specialist in the field of the theory of human origins."[35]

The physician Karl Astel, who joined the Nazi Party in 1930 and the SS in 1934, helped make the University of Jena a bastion of Nazi racial ideology. Before coming to Jena, Astel taught race hygiene (eugenics) at an SA (Nazi Stormtrooper) school in Munich, and in 1932 he helped process marriage requests for the SS by making determinations about the hereditary quality of prospective brides of SS men. Probably because of his connections with the Nazi Gauleiter (District Leader) Fritz Sauckel, Astel was appointed director of the Thuringian Office for Racial Affairs in 1933. The following year he received a professorship at the University of Jena to teach human genetics. From 1939 to 1945 he served as rector of the university. He and other Nazi leaders in Thuringia aspired to build the University of Jena into a Nazi university or even an "SS university."[36]

While in Jena, Astel held a number of Nazi Party and governmental positions, including head of the regional branch of the Racial Policy Office. His inaugural address at the University of Jena dealt with racial ideology and was published in *Nationalsozialistische Monatshefte*. This lecture was laced with Darwinian themes and explained the intersection of evolution and Nazi racial ideology. Astel claimed that one of the greatest achievements of Nazism was its recognition that humans are subject to natural laws and can thereby further biological evolution. He stated that the Nordic race had evolved through the struggle for existence and intense selection caused by the Ice Age. The harsh conditions had caused the weak to perish, leaving only the most robust to reproduce. "In this way," he explained, "through continual destruction in innumerable generations of the life that was incapable of preserving itself, through a propagation of the life that had the greatest part of those fit for life, the Nordic race and its traits, its characteristics, its talent, its superior ability of the body, mind and feelings, its camaraderie, its proverbial loyalty, and its leadership were bred."[37]

Other races, he maintained, were inferior because they had not endured as stringent a struggle. Races lacking this selective pressure were inferior, because they did not evolve as much as the Nordic race had.

Astel wrote to Heinrich Himmler in September 1937 to solicit help in recruiting Gerhard Heberer, an evolutionary anthropologist, to Jena. He told Himmler, "Our racial front would be strengthened thereby in this scientific field with an outstanding expert and a fearless fighter."[38] Himmler responded affirmatively, and Heberer received an appointment as associate professor of biology and human evolution in 1938, three years after he had become an SS officer. In the SS Heberer was assigned an honorary position in the Race Office, a subunit of the Race and Settlement Main Office; he also received an honorary position in the SS Ahnenerbe. He was promoted twice, becoming SS-Hauptsturmführer (captain) in early 1942. Heberer gave lectures on evolution to various Nazi organizations.[39] Sauckel considered this professor of human evolution so important to the Nazi cause that in 1943 he implored Bernhard Rust, the Nazi Reich Minister of Education, not to allow Heberer to be called to another university, because "I have fixed the goal of building the University of Jena into a National Socialist center of the first rank."[40] Heberer abetted Nazi racial ideology by vigorously championing Nordic racism. In a 1943 booklet he explained that the Indo-Germanic people were identical with the Nordic race, and they originated during the Ice Ages in north-central Europe, just as the human species had earlier. Heberer clearly promoted the idea that races, including the Nordic race, had evolved.[41]

Heberer was a pivotal figure in the development of the neo-Darwinian synthesis in Germany, editing what some historians now consider the most important work on evolutionary theory during the Nazi period, *Die Evolution der Organismen* (*The Evolution of Organisms*, 1943), which featured essays by leading biologists who explained the state of evolutionary theory in their specializations. The eighteen essays included discussions of paleontology, animal psychology (by Konrad Lorenz), zoology, botany, genetics, and natural selection. They also included four essays specifically on human evolution by Christian von Krogh, Wilhelm Gieseler, Reche, and Weinert. Of these, all but Reche were members of the SS.

The sheer amount of attention paid to human evolution in this textbook, and in Nazi Germany generally, is notable. The historian Thomas Junker comments in his book on the development of the neo-Darwinian synthesis in Germany that one of the key differences between those advocating the neo-Darwinian synthesis in Germany and those promoting it elsewhere is that the Germans focused much more attention on human evolution, while the Anglo-American biologists tended to ignore it.[42]

Gieseler's contribution to Heberer's anthology was an essay on "The Fossil History of Humans." His vision of evolutionary history was consistent with the newly forming neo-Darwinian synthesis, since he explained that the most important mechanisms of evolution were mutations, selection, and isolation.[43]

Gieseler, whom Junker calls one of the leading paleo-anthropologists in the world from 1930 to 1970, was appointed by the Nazi regime to a professorship at the University of Tübingen, first in 1933 as associate professor of anthropology and racial science and five years later as professor of racial biology.[44] He also served as director of the German Society for Anthropology from 1936 to 1958. Gieseler joined the Nazi Party on May 1, 1933, then joined the SA (Nazi stormtroopers) the following year, and in 1937 became an SS officer, attaining the rank of Hauptsturmführer (captain) by 1943.[45] He was affiliated with the SS Race and Settlement Main Office and also held a local leadership position in the Nazi Racial Policy Office, for whom he sometimes lectured on human evolution.[46] In 1936 Giesler wrote an entire book on human evolution: *Abstammungskunde des Menschen* (*Human Evolutionary Science*). The first volume of his work was titled *Abstammungs- und Rassenkunde des Menschen* (*Human Evolutionary and Racial Science*).

Two Objections Answered

SINCE THE Nazi reliance on Darwinian evolution becomes clear when one examines the sources, why have some historians argued that Nazis rejected Darwinian evolution? Before answering this question, I should note that relatively few historians have argued that, and many historians do understand that the Nazis accepted, and even embraced, Darwinism. Various scholars have emphasized the social Darwinism inherent in Nazi racial ideology.

One reason some historians (such as George Mosse and Peter Bowler) have erred is the mistaken belief that the Nazi insistence on hard heredity entailed a rejection of evolution. Hard heredity—championed by German biologist August Weismann—is the idea that gametes are not affected by changes to somatic cells (non-reproductive cells that make up most of your body). Weismann rejected the Lamarckian idea that organisms can evolve by passing on acquired characteristics to their progeny. The Nazis continually insisted that heredity cannot be directly affected by the environment, charging that the inheritance of acquired characteristics was a Marxist doctrine. They emphasized the continuity of racial heredity and the inability to use education or outward changes to improve heredity. The Nazis' anti-Lamarckian hard heredity is not anti-evolutionary at all, however, since Weismann was one of the leading evolutionists of the late nineteenth century. Weismann's ideas were known as neo-Darwinism in his day (though this should not be confused with the later neo-Darwinian synthesis).

When the Nazis occasionally claimed that the Nordic race had been unchanged for thousands of years, they were not claiming that it had been immutable over geologic time. Walter Gross, head of the Nazi Racial Policy Office, clarified this point in an essay on "The Racial View of History." After bashing Lamarckism, he reminded his readers that even though racial traits do not seem to change during the period of history we know (because it is too short), "selection and elimination" (a phrase often used by German evolutionary biologists to mean natural selection) do alter racial traits in longer periods of time.[47] Most Darwinists admitted that as far as we could tell, humans had not changed significantly during the course of the past several thousand years. For example, in an essay on human evolution, the evolutionary anthropologist Otto Reche argued that human races had not changed significantly in the past 20,000 to 30,000 years.[48] By rejecting Lamarckism and insisting on hard heredity, Nazi racial theorists were, in this case, in line with the best science of their day.

Another reason some historians have erred is they think the Nazis would not have wanted to affirm a common ancestor for the various human races, because a common origin would imply human equality. But this ignores the evidence, for Darwin in *The Descent of Man* pressed the case for common descent even as he embraced the view that races were unequal, not only physically, but also intellectually and even morally. Haeckel and many other Darwinists argued that evolution was evidence against human equality, not in support of it. As explained above, Lorenz, Weinert, Staemmler, and many others argued that Darwinism supports racial inequality. Indeed, Nazi racial theorists believed that the Nordic race had diverged from other races far enough in the past that it varied considerably from those other races. They also provided a selectionist account (harsh conditions during Ice Ages) to explain why it had evolved into a superior race.

Conclusion

IN SUM, the evidence of Darwin's influence on the biologists and anthropologists of Nazi Germany is wide and deep. Many of the leading anthropologists in Germany were enthusiastic Darwinists, and they were feted by the Nazis. The Nazi regime not only welcomed their Darwinian ideas, but they promoted those ideas as vital elements of Nazi racial doctrine. These anthropologists were given professorships at leading universities, and some of them participated in government committees that worked on Nazi racial policies. Some were SS officers who participated in various SS sub-organizations dealing with racial matters, such as the Race and Settlement Office. Quite a few of them regularly lectured on racial ideology for Nazi organizations and training courses. Clearly the Nazi regime approved of and advanced the careers of those professionals whose central focus was human and racial evolution, all of it rooted firmly in Darwinian evolutionary theory.

5. Nazi Eugenics and Euthanasia

THE OVERRIDING, LONG-TERM GOAL BEHIND ALL NAZI POLICIES was to promote evolutionary progress. Hitler and his Nazi comrades constantly stressed the need for the German people to promote health and biological vitality so that they—as the supposedly superior race—could triumph in the struggle for existence against other races. To help achieve this goal, in July 1933 Hitler introduced a sweeping eugenics program, which resulted in the compulsory sterilization of 350,000 to 400,000 mentally and physically disabled Germans. Under the cover of war, in late 1939 Hitler radicalized his eugenics policies by initiating a euthanasia program to kill people with disabilities. By 1945 the Nazi regime had secretly murdered about 200,000 disabled Germans, as well as tens of thousands of disabled people in occupied countries.[1]

Nazi propaganda promoting eugenics often clearly explained the Darwinian underpinnings of their eugenics ideology and policies. In 1937 the Nazi Racial Policy Office produced an educational documentary film, *All Life Is Struggle*, to woo German young people to accept eugenics. This is a silent film with written words interspersed with video. The opening lines of the film, which are matched with video of animals fighting each other, are:

> All life on the earth is struggle, struggle for existence, struggle for the preservation of the species. In the struggle for food the cowardly and unfit for life will not be satisfied. In the struggle for the female the healthy triumphs, so only the best heredity is propagated. With unyielding harshness everything is eliminated that does not meet the conditions of natural life. That is true for plants and animals—and also for humans in the same measure. The weak and unfit for life must yield

to the strong. Nature allows only the most powerful to survive. This struggle is a divine law; it serves for the perfecting of all creatures.[2]

This passage is laced with Darwinian terminology and concepts, portraying the struggle for existence as a positive force improving biological organisms. Later this film forcefully argues that humans need to conform to the laws of nature, especially the "natural law of selection." Clearly the Nazis saw eugenics as the application of Darwinian principles to human society.

After explaining that the struggle for existence eliminates the unfit and improves the species, the documentary criticizes human society for taking care of people with mental and physical disabilities, allowing them to survive and even reproduce. The film called these people derogatory names, such as "inferior," "unfit for life," "lives without value," and even "completely bestial." It informed the audience how much it costs to keep these disabled people institutionalized: "These people, who could not withstand the struggle for existence, can only be maintained at the expense of our healthy racial comrades." It closed by admonishing Germans to cultivate health and vitality, and some of the final scenes depicted German military forces marching in rank, implying that this biological vigor is also important for building German military strength.[3]

All Life Is Struggle clearly illustrates how Darwinism underpins Nazi conceptions of eugenics. However, the Darwinian influence on eugenics began long before the Nazis came on the scene.

The Darwinian Roots of Pre-Nazi Eugenics and Euthanasia

THE FOUNDING father of eugenics, Francis Galton, was profoundly impressed by *The Origin of Species*, written by his cousin Charles Darwin. In his memoirs Galton reported, "The publication in 1859 of *The Origin of Species* by Charles Darwin made a marked epoch in my own mental development, as it did in that of human thought generally."[4] In 1869 he confessed to Darwin that "the appearance of your *Origin of Species* formed a real crisis in my life; your book drove away the constraint of my old superstition as if it had been a nightmare and was the first to give me freedom of thought."[5]

In addition to helping him conquer his "superstition," Darwin's *Origin* also inspired Galton with the key idea that would shape many of his future endeavors. As he pondered the implications of Darwin's discussion of hereditary variation and evolution through natural selection, Galton concluded that humans should help shape the evolutionary process. Why not foster the prolific reproduction of those humans having the best hereditary traits, while restraining the reproduction of the inferior? Why not breed better humans, Galton wondered, so we can speed up evolutionary progress?

Galton coined the phrase "nature vs. nurture" to describe the debate between biological determinists, who believe that human behavior flows primarily from hereditary traits, and environmental determinists, who think that education and upbringing are decisive in shaping human behavior. Galton preached biological determinism, insisting in his writings, including his 1869 book *Hereditary Genius*, that not only mental abilities, but also many moral traits, are inherited biologically. Eugenics, Galton thought, should become a "new religion" that would humanely improve humans biologically. He explained, "What nature does blindly, slowly, and ruthlessly, man may do providently, quickly and kindly." However, despite his appeal to kindness, he ominously branded as enemies of the state any inferior people who dared defy reason by procreating; he suggested that such people "have forfeited all claims to kindness."[6] The state, in this case, was no presently existing state but rather Galton's vision of some utopian future state in which eugenics was law.

Darwin did not jump on Galton's eugenics bandwagon, but others found justification for such policies ready at hand in Darwin's writings, particularly in *The Descent of Man*. There Darwin remarked:

> With savages, the weak in body or mind are soon eliminated; and those that survive commonly exhibit a vigorous state of health. We civilised men, on the other hand, do our utmost to check the process of elimina-

tion; we build asylums for the imbecile, the maimed, and the sick; we institute poor-laws; and our medical men exert their utmost skill to save the life of every one to the last moment. There is reason to believe that vaccination has preserved thousands, who from a weak constitution would formerly have succumbed to small-pox. Thus the weak members of civilised societies propagate their kind. No one who has attended to the breeding of domestic animals will doubt that this must be highly injurious to the race of man. It is surprising how soon a want of care, or care wrongly directed, leads to the degeneration of a domestic race; but excepting in the case of man himself, hardly any one is so ignorant as to allow his worst animals to breed.[7]

It's important to emphasize that Darwin immediately went on from there in the book to insist that we could not "check our sympathy, if so urged by hard reason, without deterioration in the noblest part of our nature." The passage clarifies that sympathy (caring for the weak and sick), which Darwin called the "noblest part of our nature," trumps "hard reason." In the next sentence Darwin described neglecting the weak and sick as a "great present evil," indicating again that he favored helping the weak and sick. Darwin further stated, "Hence we must bear without complaining the undoubtedly bad effects of the weak surviving and propagating their kind."[8]

Darwin thereby explicitly supported helping the weak and sick and opposed harsh measures to hinder their reproduction. He claimed in the closing of that paragraph that the weak and sick marry less often and that offsets some of the negative consequences, but he himself was only willing to support voluntary measures, advising that "both sexes ought to refrain from marriage if in any marked degree inferior in body or mind."

Many of Darwin's followers, however, chose to follow the path Darwin described as "hard reason," even to the point of actively helping natural selection along. Some of them further argued that what they were doing was in fact compassionate because they thought curtailing reproduction through science was kinder than going back to the law of the jungle and letting natural selection do it.

By the 1890s and early 1900s many leading Darwinian biologists and physicians, including Darwin's son, Leonard, joined the burgeoning eugenics movement. Darwinism was a powerful influence on this movement, as many proponents explained in their writings.

In 1924 Harvard University geneticist Edward M. East vigorously promoted eugenics in his book *Man at the Crossroads*. In the preface he explained the underlying ideas motivating him to press for measures to improve human heredity: "Proper direction of human evolution is a worthy objective, a high ideal which no one should censure; for whether we will or no, our complex social organization founded for the protection of the many has set at naught many of the important factors of natural evolution."[9] In the second chapter of that book, entitled "The Biological Setting," East explained the importance of biological evolution in shaping his views. "Organic Evolution," he proclaimed, is "the greatest generalization of the human mind, a generalization which, taken with all its connotations, reaches every walk of life, modifies or justifies every custom, shows the reason for the past and points the way to the future."[10]

Non-biologists also saw eugenics as rooted in Darwinism. The birth control advocate Margaret Sanger, founder of Planned Parenthood, zealously championed eugenics. Indeed, in 1920 she insisted that birth control "is nothing more or less than the facilitation of the process of weeding out the unfit, of preventing the birth of defectives or of those who will become defectives."[11] Two years later she explained that eugenics is based on evolutionary theory: "The Eugenist points out that heredity is the great determining factor in the lives of men and women. Eugenics is the attempt to solve the problem from the biological and evolutionary point of view. Eugenics thus aims to seek out the root of our trouble, to study our humanity as a kinetic, dynamic, evolutionary organism, shifting and changing with the successive generations."[12] Sanger clearly saw eugenics and evolutionary theory as intertwined.

Galton, East, and Sanger were just a few of the many eugenics advocates who understood eugenics as a practical application of evolutionary theory to humanity. Indeed, this idea was so widespread that the 1921 International Congress of Eugenics in New York City inscribed it on certificates of appreciation sent to exhibitors. At the top of these certificates was a picture of a tree with the saying, "Eugenics is the self-direction of human evolution."[13]

The evolutionary influence on eugenics was apparent from the earliest phases of the eugenics movement in Germany, too.[14] In the 1870 edition of his major book on evolutionary theory, the leading Darwinian biologist in Germany, Ernst Haeckel, suggested that artificial selection could be used to improve human heredity. Indeed, the measure he proposed at this time—killing disabled newborns—was so radical that he could not bring himself to explicitly endorse it, though he strongly implied that he did.[15] Later, however, in 1904, he clearly stated that he approved of infanticide and confessed that he had endorsed it earlier. In 1904 Haeckel also proposed that some adults with disabilities or illnesses, whose lives he called "completely worthless," should be involuntarily euthanized.

Haeckel argued that his views about the value (or lack thereof) of human life was directly connected to his embracing of evolutionary theory. He stated, "The value of human life appears to us today, on the firm ground of evolutionary theory, in an entirely different light than it did fifty years ago."[16] It is little wonder, then, that Haeckel became an honorary president of the German Society for Racial Hygiene when it formed in 1905 (racial hygiene was a synonym for eugenics). The other honorary president of the German Eugenics Society was the prominent Darwinian biologist August Weismann.

The first work in German entirely dedicated to promoting eugenics was the physician Wilhelm Schallmayer's 1891 pamphlet *The Threatening Physical Degeneration of Civilized Peoples*. On the first page he lauded Darwin's theory for showing us that we are subject to natural selection, which brings about evolutionary progress. Then, a few pages later, he explained that his pamphlet aimed at showing which social institutions foster, and which ones hinder, the "ennobling selection of the struggle for existence."[17]

When the prestigious Krupp Prize Competition was announced in 1900, Schallmayer eagerly wrote a book addressing the assigned question: "What do we learn from the principles of biological evolution in regard to domestic political developments and legislation of states?"[18] His book-length treatment of eugenics, *Vererbung und Auslese im Lebenslauf der Völker (Heredity and Selection in the Course of Life of the Peoples)*, won the prize of 10,000 marks, a handsome sum at that time. In that book he claimed that evolutionary theory "undeniably leads" to evolutionary ethics. What would this entail? Schallmayer summed it up thus: "From all this it follows that from the standpoint of evolutionary theory, the goal of all state policy, even domestic policy, can be none other than *to prepare for the people that live in that state the most favorable conditions for victoriously surviving the struggle for existence.*"[19] In 1910 he wrote in a letter to another leading eugenicist that eugenics was indissolubly bound up with Darwinian theory.[20]

The key organizer of the German eugenics movement, Alfred Ploetz, was likewise heavily influenced by Darwinism. Ploetz founded both the first eugenics organization in the world, the Society for Racial Hygiene, and the first eugenics journal in the world. In 1892 he wrote to a friend that his main ideas about eugenics were drawn from Darwinism, and he often praised Haeckel as a key influence on his world view.[21] When he founded his eugenics journal, he told Haeckel that it would "stand on the side of Darwinism," which was also obvious from the advertisement he sent to prospective subscribers, which was saturated with Darwinian terminology and concepts.[22] One of Ploetz's co-editors was Ludwig Plate, an evolutionary biologist who later replaced Haeckel at the University of Jena. When the Nazis came to power, they appointed Ploetz and other leading German eugenicists to the Committee for Pop-

ulation and Racial Policy, which formulated the Law for the Prevention
of Hereditarily Ill Progeny, a compulsory sterilization law, which Hitler
announced in July 1933.[23] In 1936 Hitler personally bestowed the title
of honorary professor on Ploetz in recognition of his achievements in the
eugenics movement.[24]

Like the eugenics movement, the closely related euthanasia movement that preceded and influenced the Nazi program to kill people with
disabilities was powerfully influenced by Darwinism. Many of the pre-
Nazi German advocates for euthanasia—which often included involuntary euthanasia for people with disabilities—saw humans through the
lenses of Darwinian evolution.

The historians Ian Dowbiggin and Nick Kemp, in the best historical studies of the euthanasia movement in the United States and Britain
respectively, both emphasize the crucial role Darwinism played in initiating and ideologically underpinning this call for aggressive euthanasia policies. Dowbiggin states, "The most pivotal turning point in the
early history of the euthanasia movement was the coming of Darwinism to America."[25] Kemp agrees, stating, "While we should be wary of
depicting Darwin as the man responsible for ushering in a secular age
we should be similarly cautious of underestimating the importance of
evolutionary thought in relation to the questioning of the sanctity of human life."[26]

Historical studies of the German euthanasia movement have concurred with Dowbiggin and Kemp. Hans-Walter Schmuhl, a leading
historian of the German euthanasia movement, argues that eugenics
constituted an attempt to promote a new ethics based on Darwinian
science, and then euthanasia flowed from this. He explains, "By giving
up the conception of the divine image of humans under the influence of
the Darwinian theory, human life became a piece of property, which—in
contrast to the idea of a natural right to life—could be weighed against
other pieces of property."[27]

Darwinism in Nazi Eugenics

HITLER BEGAN vigorously promoting eugenics in 1923, and in *Mein Kampf* (1924–25) he explained its importance, as well as its evolutionary roots. He complained that humanitarian efforts to solve social problems were misguided, because, he thought, they violated the laws of nature. "Just as Nature does not concentrate her greatest attention in preserving what exists, but in breeding offspring to carry on the species," he wrote, "likewise, in human life, it is less important artificially to alleviate existing evil, which, in view of human nature, is ninety-nine per cent impossible, than to ensure from the start healthier channels for future evolution." Hitler then argued that social problems could only be solved by discarding policies that cause biological degeneration. In this passage he did not suggest specific eugenics measures, but he made clear his desire "brutally and ruthlessly to prune off the wild shoots and tear out the weeds."[28] Clearly Hitler saw eugenics policies as a crucial method to advance human evolution.

The official interpretation of the Nazi compulsory sterilization law was penned by Ernst Rüdin along with Arthur Gütt and Falk Ruttke, two fellow members of the Interior Ministry's Committee on Population and Racial Policy, which had crafted the legislation. Rüdin was a psychiatrist and leading figure in the eugenics movement, becoming president of the International Federation of Eugenics Organizations in 1932. When the Nazis came to power, they installed him as head of the Society for Racial Hygiene, and later the Nazis showered him with various honors, such as the Goethe Medal of Arts and Sciences in 1939.[29]

In their commentary on the eugenics sterilization law, the opening sentence of the second paragraph mentioned that the biological thought currently permeating German culture began with the work of Darwin, Mendel, and Galton, and then was fostered by Ploetz, Schallmayer, Rüdin, Fischer, Lenz, and others. Thus they explicitly connected their ideology with Darwin, and most of these other scientists they mentioned were powerfully influenced by Darwin as well.

Later in this work they explicitly discussed human evolution. They explained that civilization has contravened natural selection, which "in its effects leads the particular race to upward evolution and preservation of its hereditary health. Thus *inheritance* and *selection* are the natural managers of the racial evolution of both a people (*Volk*) and a family." However, civilization had controverted the beneficial impact of Darwinian selection by assisting the "inferior" people, such that "the reduced fitness for life, as Darwin expressed it, does not lead to elimination, but rather the effect of natural selection is transformed into its opposite through civilization and thus becomes *contraselection*."[30] Rüdin and his colleagues defended compulsory sterilization by appealing to Darwinian evolution.

In January 1934 the Nazi Interior Ministry sponsored a seminar in Munich to orient 120 psychiatrists to the new eugenic sterilization law, which went into operation that month. At that conference Gütt, an SS physician and eugenics enthusiast, delivered a speech on "The Importance of Elimination and Selection for Hereditary Health and Racial Care." He began the lecture by stating, "Nature is engaged in a biological struggle for existence!... Every species and race is purified and improved again and again through a process of selection.... As a result of biological selection the inferior individual perishes before reproducing or even before reaching the age of reproduction. Darwin called this elimination of those with reduced fitness, [and] the survival of the healthy, the result of natural selection."[31] So again we find Darwin and his theory of natural selection explicitly invoked to justify the Nazi eugenics program.

During the Third Reich, Ploetz's eugenics journal continued to publish articles about evolution, especially human evolution. In 1935, for example, Plate published an article defending evolutionary theory against anti-evolutionists. Apparently Plate had a hard time finding any anti-evolutionists among his fellow German biologists, because his article was directed against a Swedish biologist.[32]

Figure 5.1. Nazi school poster: "Elimination of the Sick and Weak in Nature." The accompanying explanation told teachers to use this to promote eugenics policies such as sterilization.

Another influential purveyor of eugenics during the Third Reich was the physician Martin Staemmler. Some of his writings were published with the explicit approval of leading Nazi officials. Staemmler joined the Nazi Party in 1931 and worked in the Nazi Racial Policy Office.[33] His Nordic racism and eugenics ideology were completely in step with Nazi ideology, so Nazi officials tapped him in 1933 to teach about genetics and eugenics in three-day physician training courses in Dresden.[34] The Nazi regime then appointed Staemmler to medical professorships at the University of Kiel in 1935 and to Breslau later the same year. Staemmler's most famous work, *Racial Care in the [German] Ethnic State*, sold over 80,000 copies during the Nazi period. In September 1935 this work was given the stamp of approval by the Nazi agency that examined books to see if they were consistent with Nazi ideology. The Nazi Party's Racial Policy Office even published some editions of this work, and the Ministry of Education recommended that school libraries acquire it.[35] During World War II the German military published an edition of Staemmler's book under a different title, *German Racial Care*, to indoctrinate their troops.[36] The 1937 copy that I examined (from the University of California's Southern Regional Library Facility) contained the stamp of ownership of the "Library of the Educational Office of the SS Main Office," so the SS apparently liked Staemmler's book, too.

Staemmler opened his book by explaining that the greatest danger facing society is its disregard for the laws of nature. He then listed four laws—*"struggle for existence, fertility, selection, heredity"*—and called them the "holiest of all laws."[37] All these laws relate to evolution, as Staemmler makes clear throughout his book. He devoted an entire chapter to "The Law of the Evolution of Living Organisms," wherein he rejected Lamarckism in favor of Darwin's theory. He even claimed that Darwin was perhaps the greatest scientist of all time. Further, he made clear that humans are included in the evolutionary process.[38] Another chapter on "The Law of Selection" explained that humans are in a struggle for existence in which "the weaker is destroyed, the stronger holds its ground."[39] Evolutionary themes pervade the book, and Staemmler appealed to evolutionary theory to defend collectivism, eugenics, and the need for living space, among other Nazi doctrines.

Staemmler's 1939 pamphlet, *Selection in the Hereditary Stream of the People*, was not only published by the official Nazi publishing house, but it had on the title page the imprimatur from the "Deputy of the Führer for the Supervision of All the Intellectual and Worldview Education and Instruction of the Nazi Party." As the title of the pamphlet indicates, selection played a central role in this work, as did the struggle for existence; one of the chapters is even entitled, "Natural Selection among Humans." Staemmler promoted the idea of human evolution in no uncertain terms: "The comparison between earlier forms and those of our days shows that in the course of nature, in the evolution of living organisms, progress occurs from lower to higher, from primitive to highly organized. He [the scientist] sees evolution, a climbing upward." He immediately added that fossil evidence demonstrated there was a time in the past when fish were the highest organisms, and only later did amphibians, reptiles, and finally humans arrive.[40] He went on to explain that evolutionary novelty arises through mutations.[41] Staemmler's views on evolution, racism, and eugenics were fully consistent with Nazi ideology, as many Nazi agencies recognized in a multitude of ways.

Darwinism also played a prominent role in the three-day courses on eugenics and racial science offered by the State Academy for Physicians' Continuing Education in Dresden, which had 4,000 participants in 1933 alone. In 1934 these courses were taken over by the newly created State Academy for Racial and Health Care, housed at the German Hygiene Museum in Dresden. The director of this academy, Ernst Wegner, who was also State Commissar for Health Care in the German state of Saxony, published some of the lectures as *Racial Hygiene for Everyone* (1934). After an opening chapter by Wegner on the effect of race on history, the anthropologist Otto Reche gave an entire lecture on the "Origin of Humans and Its Races." Reche explained the prevailing view among anthropologists that humans had arisen from primate ancestors. He also claimed that the Ice Ages played a formative role in human evolution. Nonetheless, he argued for a selectionist model of evolution, insisting that the environment played only an indirect role. For those wanting to explore these issues of eugenics and racial science further, Wegner recommended reading the works of Günther, Staemmler, and the Baur-Fischer-Lenz text, all of which contained Darwinian elements (as we have seen).[42]

Another important work demonstrating the close links between Darwinism and eugenics was *Hereditary Science, Racial Care, and Population Policy: A Question of the Destiny of the German People*, written by Alfred Kühn, Martin Staemmler, and Friedrich Burgdörfer. This book carried the imprimatur of the Nazi Party.

Kühn had been a professor of zoology and genetics at the University of Göttingen since 1920. In his section on hereditary science he devoted an entire chapter to human evolution. He considered evolution, including human evolution, a fact supported by overwhelming evidence. He believed that mutations and natural selection drove the evolutionary process. He rejected Lamarckism and asserted, "The progress of research has substantiated the view of C. Darwin and A. Weismann, that for the formation of species natural selection is of the greatest importance."[43] German biologists and physicians who promoted eugenics during the Third Reich overwhelmingly believed in human evolution via natural selection and considered it inextricably linked to eugenics.

Darwinism in Nazi Films Promoting Eugenics and Euthanasia

THE NAZI film *All Life Is Struggle*, discussed in this chapter's introduction, was only one among several propaganda films to feature Darwinian themes. The historian Michael Burleigh asserts that a common theme of Nazi documentaries about eugenics and euthanasia was "a crudely Social Darwinian view of life as a perpetual struggle for survival amidst a hostile natural environment."[44] The historian Ulf Schmidt concurs, stating that the Nazi Racial Policy Office's documentaries "referred to the social Darwinian ideology of the continuous struggle for survival in human society, hereditary health and race hygiene."[45]

The 1936 silent documentary *Hereditarily Ill* was one of those films produced by the Nazi Racial Policy Office. It depicts the horrors of hereditary mental illnesses, along with the huge expense of caring for people with mental disabilities. One of the key points of this film is that society violates the laws of nature by setting aside the struggle for existence and tending the mentally ill: "In the struggle for existence only the healthy survive. [Today] what would perish according to the laws of nature in a state of freedom is protected and cared for." Caring for the hereditarily ill has led, according to this film, to a dramatic increase in the numbers of people with mental illnesses, whom it disparages as "lower than animals." Preventing these people from reproducing is a "moral command" in line with the laws of nature.[46]

Burleigh, who wrote the script for the documentary *Selling Murder: The Secret Propaganda Films of the Third Reich*, states that the 1935 Nazi film *The Inheritance* "remorselessly advocates killing the weak." *The Inheritance* explicitly appeals to nature, especially natural selection and the struggle for existence, to justify not only eugenics, but also killing people with disabilities.[47] The Nazi film blended together a plot line with a documentary by showing professors and lab assistants making nature films showing animals fighting. One of the characters stated, "We made a series of film clips about the struggle for existence." They then watched the documentary they were making, which included this narration: "Among these animals, nature does its merciless selection. The sick bird falls prey to the cat. The weak hare is caught by its hunter. Even members of the same species have their fights in which the weaker is the victim. Some female birds have the instinct to eliminate their poorly developed offspring."

This last statement is, of course, ominous, in that it implied that nature sanctions the killing of the weak of one's own species. Even more ominously, upon hearing this statement about the birds killing their own offspring, one of the lab assistants interjected, "So the animals pursue a proper racial policy?" The professor replied, "In certain respects, yes." The film then juxtaposed images of mentally ill people in insane asylums with scenes from nature. Burleigh surely summarizes the Nazi film's central thesis accurately when he has the film's narrator say, "The message is clear: The so-called weak should be eliminated."[48]

These earliest Nazi films were somewhat amateurish ventures probably intended for schools or Hitler Youth meetings. Hitler, who loved movies himself, then ordered the production of a more sophisticated film on eugenics, so the Nazi Racial Policy Office enlisted professional filmmakers. The Nazi regime mandated the screening of the resulting documentary, *Victims of the Past*, in all the movie theaters throughout

Germany. This film, like *The Inheritance*, explicitly appealed to Darwinian themes. "All life on this earth is a struggle for existence," it proclaimed, and then it continued,

> Everything weak unfailingly perishes in nature. We have sinned terribly against this law of natural selection in the last decades. We have not only preserved the life [of the weak], but we have even allowed them to reproduce. All this misery could have been prevented, if we had previously prevented the reproduction of the hereditarily ill. The prevention of hereditarily ill progeny is a moral command. It means practical love of one's neighbor and the highest respect for the God-given laws of nature. Whoever prevents weeds, assists the valuable ones.[49]

After featuring many film clips showing people with hereditary mental illnesses, it lamented that "these poor beings, however, are a caricature of life. We have set aside the law of selection through the struggle for existence. But we have given these people the possibility to double and to multiply their bodies in their children." Once again, this film presented the Darwinian struggle for existence as a beneficent force, while negatively portraying the expense and "burden" of caring for the weak and sick.

In 1939 Hitler authorized the killing of people with disabilities in a secret operation codenamed T-4. The directors of the T-4 program recognized that their activities, which were still technically illegal, faced opposition from significant segments of the German public. Thus, they recruited leading film directors and actors to produce a feature film about euthanasia entitled *I Accuse*, which was released in 1941. It was viewed by over fifteen million Germans.[50] The primary plot dealt with a woman's desire to die when she was diagnosed with multiple sclerosis and the resulting murder trial, when her physician husband helped her commit suicide. Although the main story was about the woman's voluntary euthanasia, a sub-plot raised the issue of involuntary euthanasia in a sympathetic way. One of the characters stated, "I've never really considered whether one ought or ought not to allow people to die. But what must be right for an animal must be right for a human being. In the natural world... there is a remorseless law. Everything that is weak is overwhelmed and destroyed by the strong. Everything weak dies and everything strong stays alive."[51] This reiterated the Darwinian theme that pervaded eugenics discourse during the Third Reich.

Conclusion

EUGENICS IDEOLOGY, both before and during the Third Reich, relied heavily on Darwinian biology. Eugenicists stressed that the Darwinian process of natural selection through the struggle for existence improved species. And they lamented that modern policies and institutions stymied natural selection, leading to biological decline instead of evolutionary progress. To set things right they proposed measures to artificially restrict the "unfit" or "inferior" from reproducing. Some more radical eugenics proponents even suggested killing people with mental disabilities, whom they branded as "lives unworthy of life." The Nazi regime was clearly in the camp of these radicals, killing over 200,000 people with disabilities in less than five years.

6. Darwinism in Nazi Propaganda

The official Nazi Party newspaper, *Völkischer Beobachter*, focused mostly on politics and current events and only rarely mentioned scientific issues. Nonetheless, occasionally it featured articles that honored Charles Darwin or Ernst Haeckel for their contributions to evolutionary theory. In 1932 the *Völkischer Beobachter* published an article simply entitled "Darwin," which claimed that Darwin's theory was the theoretical foundation for eugenics and racial theory, which, of course, were central features of the Nazi worldview. The article, written by an anonymous professor, explained that evolution was a well-established scientific truth that was not debatable, and that Darwin's theory of natural selection had triumphed over Lamarckian theory. It called on fellow Germans to honor "the great scientist and scholar Charles Darwin."[1]

As we shall see in this chapter, official Nazi periodicals, science journals, and official Nazi propaganda pamphlets written to propagate Nazi racial ideology all concurred with this perspective. They honored Darwin and championed his theory of natural selection, rejecting the Lamarckian theory of evolution. Whenever they overtly discussed evolution, they proclaimed it as a scientific fact and rejected any attempts to question it. Further, they portrayed biological evolution as an important factor in their worldview.

Evolutionary Theory in Official Nazi Periodicals

The *Völkischer Beobachter* was just one of many Nazi periodicals to laud Darwin and his theory. Various Nazi periodicals featured articles positively discussing evolution, including human evolution and its relationship to racial theory. Some of these articles explicitly attacked anti-evolutionary thought. The anti-evolutionary ideas they opposed, moreover, were not being published by fellow Nazis, but instead were being promoted by religious periodicals, especially the Catholic periodical *Natur und Kultur*. In the course of my research, I have surveyed quite a few Nazi periodicals, and I have never discovered a single article in them attacking or even calling into question evolutionary theory. Some articles argued over the details of evolutionary theory, and they might even criticize Darwinism as too individualistic. However, these articles always embraced the common descent of organisms, and the vast majority taught the Darwinian mechanism of natural selection through the struggle for existence. They also consistently espoused the ideas of human evolution and the evolution of races, and argued that this evolutionary framework buttressed the Nazi views on racial inequality and racial competition.

In 1934, on the occasion of the evolutionary biologist Ernst Haeckel's hundredth birthday, the *Völkischer Beobachter* ran a story about Haeckel by the evolutionary biologist Viktor Franz. While Franz expressed some criticisms of Haeckel's combative anti-Christian stance, he lauded Haeckel for his contributions to evolutionary biology.[2] In 1939, on the twentieth anniversary of Haeckel's death, *Völkischer Beobachter* carried an article even more laudatory toward Haeckel. It not only applauded his contribution toward evolutionary biology, but also highlighted Haeckel's promotion of human evolution as a worthy achievement.[3] These articles fully supported evolutionary theory, including human evolution, and presented Darwinism as an important foundation for other elements of Nazi ideology, such as racial theory and eugenics. The view among some people today that human evolution is incompatible with Nazi racial ideology was apparently not a view shared by the *Völkischer Beobachter*.

Another important official Nazi Party publication, *Nationalsozialistische Monatshefte*, was edited by Alfred Rosenberg, a leading Nazi ideologist. This monthly journal occasionally featured articles promoting evolution (including Karl Astel's lecture discussed in Chapter 4). In a 1935 article the botanist Heinz Brücher praised Haeckel for paving the way for the Nazi regime. In addition to mentioning Haeckel's support for eugenics and euthanasia, Brücher highlighted Haeckel's role in theorizing about human evolution. Brücher reminded his readers that Haeckel's view of human evolution led him to reject human equality and socialism.[4] In 1941 Brücher published another article in *Nationalsozialistische Monatshefte* on evolution through natural selection. Several times he stressed that the principles of evolution were valid for humans, just as they were for other organisms. He closed the essay by explaining the practical implications of evolutionary theory. "The hereditary health of the German *Volk* and of the Nordic-Germanic race that unites it must under all circumstances remain intact," he wrote. "Through an appropriate *compliance with the laws of nature*, through *selection* and planned *racial care* it can even be increased. The racial superiority achieved thereby secures for our *Volk* in the harsh *struggle for existence* an advantage, which will make us unconquerable."[5] In Brücher's view, then, evolutionary thinking was an essential ingredient of racial ideology, not a hindrance to it.

In 1936 the evolutionary biologist Gerhard Heberer launched an attack on anti-evolutionists in *Nationalsozialistische Monatshefte*. He began the article by lauding Haeckel and stressing the affinities of Darwinism with Nazi ideology. He argued that hereditary principles and evolutionary theory were inseparable. He offered evidence for evolution, including human evolution, and criticized the Catholic journal *Natur und Kultur* for opposing evolution.[6]

Another official magazine published by the Nazi Party, *Der Schulungsbrief*, carried an article in 1939 explaining how evolution fit into Nazi racial ideology. In his essay on "Biology and Worldview," Theodor Arzt focused almost exclusively on evolutionary biology. He stated, "We see that biology and its conclusions represent essential foundations for our worldview," since the laws of nature apply to humans as well as plants and animals. He also explained that humans, especially the Nordic race,

Figure 6.1. Nazi periodical cover with the message "Life Requires Struggle."

evolved in large part because they endured harsh conditions. Humans, he claimed, did not originate in a paradise, but in north-central Europe during the Ice Age.[7]

The Hitler Youth leader Baldur von Schirach also promoted evolutionary theory in the Hitler Youth journal he edited, *Wille und Macht.* In 1936, on the sixty-fifth anniversary of the birth of the race theorist Ludwig Woltmann, this journal carried an article and book review on Woltmann that made clear its commitment to evolutionary theory. In the article Ernst Lange correctly portrayed Woltmann as a synthesizer of the evolutionary theory of Darwin and Haeckel with the racial ideology of Gobineau and Houston Stewart Chamberlain. He then called Darwin and Haeckel the fathers of modern biological thought, on which racial science was built. He stated that "today for the first time the great scientists [Darwin and Haeckel] celebrate their true victory, since an entire Reich has begun to arrange its highest principles according to the findings of biological research."[8] Thus Lange argued that the "highest principles" of Nazism were taken from Darwinian biology.

If we want to find out what the official Nazi position was on racial matters, it would be useful to examine the position of three key racial journals: *Neues Volk,* the official publication of the Nazi Party's Racial Policy Office; *Volk und Rasse,* published by the Reich Committee for the People's Health and the German Society for Racial Hygiene; and *Rasse: Monatsschrift der Nordischen Bewegung. Neues Volk* was a glossy popular magazine filled with brief articles and many pictures dealing with human heredity, the disabled, eugenics, twin research, racial mixture, black people, dying races, Gypsies, having large families, and related topics. Many of these articles were impregnated with evolutionary themes, such as selection and the struggle for existence, though I did not notice any articles devoted entirely to evolutionary theory. However, in many issues the editors compiled lists of recommended books on racial themes. Some of these books focused on human evolution, such as two by the evolutionary anthropologist Hans Weinert.[9] In recommending Ludwig Woltmann's book *Political Anthropology,* which explained how to apply Darwinism to public policy, the editor of *Neues Volk* stated that Woltmann "is rightly named a forerunner of the völkisch and racial thought, thus the worldview that became the foundation of National Socialism."[10]

Neues Volk also reported on a 1938 training course for officials in the Racial Policy Office, during which the evolutionary anthropologist Wilhelm Gieseler lectured on "The Evolutionary Descent of Humans."[11] Clearly, then, the Racial Policy Office saw human evolution as an integral part of their racial ideology.

While the Racial Policy Office's stance on evolution and racial theory comes through pretty clearly from examining *Neus Volk*, its official publication, its stance becomes even clearer when we examine the views of its head, Walter Gross. In 1943 Gross published an article in *Die Naturwissenschaften* (*The Natural Sciences*) discussing various debates about how evolution occurs. At the beginning of the article, he affirmed that evolutionary theory (*Deszendenzlehre*) is "one of the best-established theories of natural science." He then explained that based on modern genetics, scientists have rejected Lamarckism. Rather, he insisted, evolution proceeds by mutation and natural selection, what would later be called the neo-Darwinian synthesis. The main thrust of his article was to take issue with a few German scientists who were proposing rapid evolutionary changes to account for gaps in the paleontological record. Gross believed that the gaps in the fossil record were only gaps in knowledge that would likely be filled in later.[12]

Volk und Rasse was another official Nazi publication on racial theory, but it was more academic, containing longer articles by anthropologists. In July 1933 the anthropologist Bruno Kurt Schulz became sole editor (he had been co-editor earlier), and some leading Nazi officials placed their names on the masthead: Heinrich Himmler, Richard-Walther Darré (Minister of Agriculture and head of the SS Race and Settlement Main Office), Arthur Gütt (a high-ranking Interior Ministry official in charge of racial policy), and Ernst Rüdin (discussed in Chapter 5). In 1932 (thus before the Nazis came to power) Schulz had joined the Nazi Party, the SS, and the Race and Settlement Main Office. According to

Isabel Heinemann, Schulz became "one of the central figures" of the Race and Settlement Main Office, leading its Race Office from 1932 to 1935 and again from 1942 to 1944. In 1942 the Nazi regime appointed Schulz professor of anthropology at Charles University in Prague, so he could supervise racial screening in the Protectorate of Bohemia and Moravia.[13]

Under Schulz's editorship—and with Himmler's and Darré's names on the masthead—*Volk und Rasse* published several articles entirely devoted to human and racial evolution. Heberer wrote numerous articles for *Volk und Rasse*, many of them combating creationism. In 1937, for instance, Heberer published a two-part article lambasting the "ultra-montane-Jesuitical" forces that opposed human evolution. In the article he discussed human fossil finds as evidence for human evolution. He argued in this article that evolution "forms the foundation for racial history. Races are something evolving and every consideration of race leads of necessity to the question of its origin and its evolution."[14]

Eugen Fischer, director of the Kaiser Wilhelm Institute for Anthropology, Human Heredity, and Eugenics, also published an article on human evolution in Schulz's journal. In his essay, "The Origins of Human Races," Fischer stated that races who faced squarely "the pitiless struggle causing selection" in harsher environments had "bred the highest mental characteristics." This was true of the "Nordic race in the harsh struggle for existence at the border of the Ice Age's glaciers."[15]

Volk und Rasse also carried a brief article in 1938 by Christian von Krogh describing a new display on "The Evolutionary and Racial Science of Humans" in Munich. Krogh, who contributed an essay on human evolution to Heberer's *Evolution of Organisms*, was an avid Nazi, joining the party in 1930 and later serving as an SS officer in the Security Service (SD) in France during World War II.[16] He received his doctorate in anthropology at the University of Munich in 1935 under Theodor Mollison, the organizer of this 1938 exhibition. After receiving his doctorate Krogh worked as Mollison's assistant, so Krogh probably helped work on this display, which he claimed was the first permanent exhibit of its kind. The exhibit had two sections: evolutionary history and racial science. Krogh reported that the opening ceremony on April 2, 1938, was attended by dignitaries of the state and party. This, he thought, was only fitting, since "with the foundational importance of the natural history of humans in our National Socialist worldview, it is right that it demands special attention."[17] The perception from inside the Third Reich, then, was that evolutionary theory was not only compatible with, but integral to, Nazi racial ideology.

Discussions of human evolution in *Rasse*, the monthly journal of the Nordic Ring, reinforce this conclusion. In the first volume of *Rasse* in 1934, Weinert wrote an article, "On the Primitive History of the Nordic and Phaelic Race," in which he taught that humans had evolved from primates to ape-men to Pithecanthropus to Neanderthal to Cro-Magnon to modern humans.[18] Also in the first volume were two articles by evolutionary biologists defending Lamarckian evolutionary processes, and an article wherein Günther expressed belief in human evolution.[19] In the second volume in 1935 Kurt Holler attacked Lamarckism, arguing that evolution proceeds by mutations and selection without Lamarckian mechanisms.[20] Holler's position would prevail in Nazi Germany, as both the majority of scientists and Nazi Party officials rejected Lamarckism in favor of Weismann's insistence on hard heredity. Not only did other articles in *Rasse* promote human evolution (and none that I found opposed it), but some articles praised Woltmann and Georges Vacher de Lapouge—both synthesizers of Darwin and Gobineau—as forerunners of the Nordic movement.[21]

Scientific Journals during the Third Reich

DURING THE Third Reich the scientific journal *Zeitschrift für die gesamte Naturwissenschaft* (*Journal for All the Natural Sciences*) published various articles discussing and debating evolutionary theory. The historian Robert Richards, in his book *Was Hitler a Darwinian?*, pounces on two of these articles in 1937–38 by Kurt Hildebrandt and G. Hecht and insists that they were anti-Darwinian. Richards wrongly claims that this demonstrates that the Nazis did not embrace human evolution. However, such a claim ignores the mountains of contrary examples of Nazi biologists, Nazi policy makers, and Nazi writers explicitly promoting evolutionary theory in prominent venues. It also ignores the fact that neither Hildebrandt nor Hecht denied that evolution occurred; they were simply debating evolutionary mechanisms.[22] Both were full-fledged evolutionists. Hecht, for instance, stated in his article, "The fundamental idea of evolution—according to which every living thing on our earth descends from predecessors and ancestors, and in the course of the earths' history gradually grows from older and simpler forms—is founded on a reality that is totally uncontroversial today." Later in the article he claimed, "Viewed today from the standpoint of worldview and science the age of Darwin and Haeckel in biology was essentially the age of the final triumph of the (correct) theory of evolution and descent, in which both scientists played a decisive role."[23] On two counts then, either alone decisive, Richards' use of these articles as evidence that German biologists did not embrace evolution is wildly mistaken.

Further, an interesting exchange also took place in this same journal in 1940, when M. Westenhöfer, a professor of anatomy at the University of Berlin, wrote an article claiming that humans did not evolve from primates. Richards again proclaims that this article is more evidence for his position.[24] However, Westenhöfer's essay was not anti-evolutionary at all, because he was arguing that the last common ancestor of humans and apes was not a primate, but some more primitive mammal. He was arguing *for* human evolution, not against it, though his vision of evolution was rather unorthodox and did not gain wide acceptance among biologists.[25] In the same journal Krogh published a rebuttal to Westenhöfer's article, defending the views of Haeckel, Mollison, and most other biologists and anthropologists, who believed that humans did evolve from simians.[26] The journal's editor, Ernst Bergdolt, closed the discussion with an article that seemed to favor Krogh's position.[27] So,

Westenhöfer, Krogh, and Bergdolt, as well as Hildebrandt and Hecht, all accepted evolution, though they may have disagreed about some details regarding how it happened. Despite Richards' claims, these articles are actually evidence against his position.

Importance of Evolution in Nazi Racial Propaganda

To DEMONSTRATE the important role Darwinism played in Nazi racial propaganda, I will examine three sources: the SS pamphlet *Racial Policy*, the *Curriculum for the Worldview Training in the SS and Police*, and a propaganda pamphlet written for the German military during World War II, *What Are We Fighting For?* All of these sources explicitly endorsed Darwinian evolution, including (and especially!) the evolution of humans. Further, these sources do not just mention Darwinism in passing, but accord it a prominent place in Nazi racial ideology.

Darwinism played a central role in the anonymously written SS propaganda pamphlet *Racial Policy*. As indicated on the final page, where the material is divided into eleven class periods, this pamphlet was used for training in Nazi ideology. The opening pages explained that the central concepts underlying racial ideology are hard heredity and racial inequality. The pamphlet authors claimed that racial inequality has come about because evolution proceeds by struggle. Different races simply do not evolve at the same pace, so they are at different levels. The authors then asserted that what they regard as the three main human races— European, Mongolian, and Negro—were subspecies that branched off from a common ancestor about 100,000 years ago. They argued that races evolved through selection and elimination, and the Nordic race became superior because it had to struggle in especially harsh conditions. Throughout this pamphlet the terms "higher evolution," "struggle for existence," and "selection" are core concepts that occur repeatedly. The pamphlet also mentions mutations as a source of evolutionary novelty.[28]

In a section of the book entitled "The Purpose of Life," the authors explained, "To preserve and multiply oneself is the deepest purpose of life... The preservation and multiplication of life, however, includes the drive for improvement, for higher evolution and perfection, which exists within all life.... As species arise and perish, the all-encompassing life on the earth takes on ever new forms with the goal of growing perfection of the individual and the species, its higher evolution, and the improvement of functions."[29] The statement exudes a teleology that runs contrary to some conceptions of Darwinism, but it nonetheless demonstrates the importance of human evolution in Nazi ideology.

Another training manual aimed at inculcating a Nazi worldview in SS members was *Curriculum for the Worldview Training in the SS and Police*, written sometime after 1941. The training course's fourth and final section was "The Laws of Life Foundational to Our World View." One of its eight classes was entirely on evolution and insisted that the evidence for evolution—including the embryonic recapitulation of one's ancestry (an idea that Haeckel had put forward) and the role of mutations in altering species—was overwhelming. A different class period, a subsection on "Struggle for Existence (Selection)," taught that humans arose through the struggle for existence in the Ice Ages.[30] SS training, then, included significant doses of teaching about human evolution.

Finally, another important pamphlet for Nazi ideological indoctrination was *What Are We Fighting For?*, a work published by the German military in 1944. Hitler wrote a letter approving of the booklet, asking officers not only to read it for their own instruction, but also to use it in ideological training sessions for their troops.[31] This pamphlet repeatedly stressed the importance of natural selection and the racial struggle for existence to preserve and improve the human species. The Nazi commitment to "higher evolution" of humans is a major theme. The authors argued that the Nordic race was already the highest evolved race, but they aimed at improving the human species yet more. "The means to produce this new human type," they asserted, "is instruction managed in the spirit of the National Socialist worldview. The precondition for it is maintaining purity [of blood] and advancing the evolution of our blood through breeding."[32]

Whoever wrote this pamphlet apparently did not see any conflict between human evolution and Nazi racial ideology, for they stated, "Our racial thought is only the 'expression of a worldview,' which recognizes in the higher evolution of humans a divine command."[33] The notion that advancing human evolution is a divine mandate was not unusual in Nazi ideology and reflected either a pantheistic or deistic notion of a divinity employing naturalistic processes, including the struggle for existence, to produce ongoing evolutionary progress.[34]

Objection: The Nazis Just Used Darwinism as a Propaganda Tool

IT's UNDENIABLE, then, that the Nazis employed Darwinism widely in their propaganda efforts. But one might object that the Nazi use of Darwinism was purely a rhetorical strategy: That is, Hitler and other Nazis shrewdly co-opted a dominant scientific paradigm in the service of their insidious political goals. And if there had been another dominant scientific view of origins, they would have claimed that for their cause. There are multiple problems with this objection.

First, it seems to rest on the false assumption that the Nazi regime was primarily opportunistic. Most historians today recognize that Hitler and most leading Nazis were not primarily opportunists but, rather, fanatical ideologues. Sure, Hitler and his comrades were willing to lie to the public if it brought political advantage. However, those lies were always to try to advance their heartfelt ideology, not just to attain power for power's sake.

Second, we have considerable evidence that Hitler and leading Nazis did not just use Darwinism for public consumption, but promoted it in private conversations. It was not just a superficial add-on to gain support for unrelated ideas and policies.

Third, and probably most importantly, this objection fails to recognize that leading Darwinian biologists and anthropologists were promoting scientific racism in the pre-Nazi period. The Nazis were influenced by this scientific racism. Darwinism was an essential part of Nazi racial ideology from the start. It is not like Nazis had their racist ideology in place, and then added Darwinism to the mix to gain more public support. Racism and Darwinism were closely aligned long before the Nazis developed their ideology.

Conclusion

NOT ONLY science journals, but also the most important Nazi periodicals, along with pamphlets written to teach the Nazi worldview, all taught the importance of evolutionary biology in Nazi ideology. The authors considered human evolution especially important, because they believed it supported their vision of racial inequality and racial struggle, fundamental parts of the Nazi worldview. No Nazi journal or official Nazi publication (at least, of which I am aware) published articles or essays denying human evolution. However, some did publish essays bashing creationism and anti-evolutionary ideas. Though there was some debate about the exact way that evolution occurred, the version of evolutionary theory most Nazis preferred was the Darwinian theory of natural selection through the struggle for existence.

7. DARWINIST ERNST HAECKEL IN THE THIRD REICH

IN THE ANGLOSPHERE TODAY, IF ONE KNOWS ANYTHING ABOUT THE German biologist Ernst Haeckel (1834–1919) it is that he was an early and vigorous proponent of Darwinism. If one knows one other thing about him it may be that he produced a set of embryo drawings of various species that greatly exaggerated the similarities among the embryos, apparently the better to cinch the case for Darwin's idea of common descent. Haeckel's drawings so distorted the appearance of the embryos that even leading proponents of evolutionary theory eventually distanced themselves from them, even as the drawings long remained a popular tool for promoting evolutionary theory in introductory treatments of the theory.[1] If the Nazis were pro-Darwinian, then one would expect them to have embraced the evolutionary thinking of such a zealously pro-Darwinian German biologist. But at least two historians argue otherwise.[2]

Historian Robert Richards has done everything possible to distance Haeckel from Nazism. And while historian Daniel Gasman has argued vociferously that Haeckel was the key progenitor of Nazi ideology, Gasman rejects the idea that Haeckel's evolutionary thinking was attractive to the Nazis. Part of the problem with both Gasman's and Richards's analyses is that both studied Haeckel's own position in detail, but only cursorily examined materials from the Nazi period. What is missing from their work is an adequate exploration of the Nazis' own discussions about Haeckel.[3] What did the Nazis themselves think and say about Haeckel? Did they see Haeckel as a precursor, or did they reject his ideas?

Gasman stresses the Nazi affinities with Haeckel, but he admits that during the Third Reich Haeckel never reached the iconic status of other shapers of Nazi ideology. However, because Gasman never investigated in detail the debate about Haeckel during the Third Reich, he incorrectly assumed the reason many Nazis were unenthusiastic about Haeckel was that they were "suspicious of the idea of human evolution."[4] I have already demonstrated that Nazis did indeed believe in human evolution, so Gasman is mistaken. The reason for the lack of enthusiasm must lie elsewhere.

Richards, on the other hand, claims that Haeckel "was initially recruited to the side of National Socialism but then quickly rejected by party functionaries." This brief "recruitment phase" ended in 1937, according to Richards, when Günther Hecht, an official in the Nazi Racial Policy Office, issued what Richards calls a "quasi-official" article opposing Haeckel.[5] In *Was Hitler a Darwinian?* Richards goes even further. "It is quite clear that Darwinian evolutionary theory held no special place within the community of biologists supportive of National Socialism," Richards asserts. "Rather, biologists and philosophers most closely identified with the goals of the Nazi Party and officials in that party utterly rejected Darwinian theory, especially as advanced by Darwin's disciple Ernst Haeckel."[6]

The preceding chapters show how erroneous this claim is in relation to Darwinism. Now let's consider the problems with the Haeckel side of the claim. Contra both Gasman and Richards, Nazi biologists and most Nazi leaders—even those who criticized Haeckel—accepted the biological theory of human evolution, though they did not always agree on the details. They might disagree, for instance, on what the exact evolutionary precursors were of today's humans, or on what role is played by the inheritance of acquired characteristics, or on whether evolution is gradual or proceeds by rapid changes. But almost all Nazi biologists and Nazi leaders agreed that humans had evolved.

In examining the internal debates among Nazi evolutionists, what we find is that Haeckel was almost universally praised—even by many of his critics—for his contributions to evolutionary biology, and that he anticipated and contributed to the development of Nazi ideology. In his 1936 biography of Haeckel, the biologist Heinz Brücher pointed to several important features of Haeckel's thought that presaged Nazi ideals: 1) biological racism; 2) eugenics; and 3) euthanasia for disabled infants. Remarkably, several years before the Nazis began their secret T-4 "euthanasia" program, Heinz Brücher was publicly lauding Haeckel for advocating the killing of disabled people.[7] This is striking, because most Nazis, including Hitler, knew that killing people with disabilities was too radical for most Germans, so they usually either kept quiet about it or else promoted it only very subtly in eugenics propaganda. Others during the Nazi period praised Haeckel for rejecting socialism and human equality.

However, Haeckel was never honored by Nazi writers for being a great anti-Semitic thinker. Even the most glowingly positive tributes to Haeckel in the Third Reich did not portray Haeckel as the most important progenitor of Nazi ideology. So how is it that the German biologist received such high praise from Nazis for his evolutionary ideas, but was generally passed over in this other regard? The key to this disjunction seems to rest in an institution Haeckel was long associated with, the Monist League.

Organization non Grata

As SOON as the Nazis came to power, they dissolved the Monist League, the organization Haeckel founded in 1906 to propagate his monistic worldview. The reason for this was pretty simple: the Monist League in the 1920s and early 1930s supported pacifism, feminism, and leftist politics.[8] Further, it issued explicitly anti-Nazi political statements in 1932–33, just before the Nazis took over. In 1932 the Monist League joined other freethinking organizations by signing a political declaration that attacked fascism and the "enemies of freedom and democracy."[9] In early 1933, just prior to the Nazi seizure of power, the Monist League's journal announced that it was opposed to fascist barbarism and supported "freedom, peace, and socialism."[10] By positioning itself as overtly anti-Nazi, the fate of the Monist League was sealed once the Nazis had sufficient power to squelch it.

It is unlikely that the Nazi regime disbanded the Monist League because of its association with Haeckel or because they disagreed with monistic metaphysics. No one, as far as I know, was ever persecuted in Nazi Germany simply on the basis of views about metaphysics.[11] Academic philosophers during the Third Reich were divided into two main camps: idealists, some of whom regarded the German philosopher Johann Gottlieb Fichte as a forerunner of Nazi ideology, and existentialists, some of whom suggested that Nietzsche was the forebear of Nazism.[12] The Nazi Party tolerated many perspectives on metaphysics, and many leading Nazis, such as Martin Bormann and Hitler himself, upheld views very close to Haeckel's pantheistic monism.[13] Hitler even warned his comrades against getting enmeshed in battles over metaphysics, because, he explained, Nazism rested on science, and science at that time had not achieved knowledge about metaphysics. His opposition to the Christian churches, Hitler explained, had nothing to do with their metaphysics.[14] Because the Nazi regime was trying to avoid battles over metaphysics to concentrate on their political objectives, it is unlikely that it dissolved the Monist League because of concern about Haeckel's metaphysics.

In any case, Heinrich Schmidt, the director of the Ernst Haeckel House in Jena and a leading figure in the Monist League, did not interpret the demise of the Monist League as a prohibition on pro-Haeckel advocacy. Schmidt was editor of the Monist League's journal from 1930 until it ended publication in February 1933. That same year he began publishing a more politically circumspect journal, *Natur und Geist*, which continued promoting a monistic worldview at least until 1938.[15]

Further, in early 1934 Schmidt celebrated the hundredth anniversary of Haeckel's birth by publishing a laudatory biography. In the foreword, written in January 1934, Schmidt linked Haeckel to the current regime, stating, "In the new Reich his ideas about biology and the interconnectedness of nature are celebrating a surprisingly powerful resurrection. The religious trajectory of the present is often travelling in the course of his simple, yet sublime, nature religion."[16] Thus Schmidt suggested that even after the dissolution of the Monist League, Haeckel's pantheistic metaphysics still resonated with the new regime.

Despite dissolving the Monist League, the Nazi regime left the Ernst Haeckel House, Haeckel's former residence that had been transformed into an archive and museum, unscathed. When Schmidt died in May 1935, Nazi Minister of Education Bernhard Rust named Viktor Franz, an associate professor of evolutionary biology at the University of Jena and a devoted follower of Haeckel, the new director of the Ernst Haeckel House. In 1936 the Ministry of Education promoted Franz, who had joined the Nazi Party in 1930, to full professor; he delivered his inaugural lecture on evolutionary theory. Not only did he regularly lecture at the University of Jena on evolutionary biology, but in the winter semester of 1938–39 he even offered a course on "Haeckel's Life and Work in the Light of the Present."

In May 1937, when Franz called for a celebration at the Ernst Haeckel House after two years of renovations, the Thuringian governor's office and the Thuringian Office for Racial Affairs both sent representatives to speak at the gala event.[17] In 1939 the Nazi Minister of Education improved the status of the Ernst Haeckel House by granting it the title "Institute for the History of Zoology, especially the History of Evolutionary Theory." This gave the Ernst Haeckel House the right to supervise and grant doctoral degrees. These developments led Franz to conclude that Haeckel's prestige was rising in Nazi Germany. In 1939 he stated, "For a long time it was considered good form to give Haeckel the cold shoulder. Today all the slander is dropping away from him."[18]

The New Ernst Haeckel Society, Founded in 1942

ANOTHER POWERFUL indication that the end of the Monist League was not a sign of Nazi hostility toward Haeckel himself, much less his evolutionary views, came in 1942, five years after Richards claims Haeckel was blackballed by the Nazi regime. In that year Haeckel's admirers formed a new organization to promote Haeckel's achievements, and this organization lasted until the demise of Nazism. Formal discussions about forming the Ernst Haeckel Society began in July 1941, when evolutionary biology professor Gerhard Heberer approached fellow SS officer Karl Astel, the rector of the University of Jena, to seek his support. Astel, a longtime fan of Haeckel who specialized in human genetics, was trying to turn the University of Jena into an "SS university." When Brücher published his biography of Haeckel in 1936, Astel wrote the foreword, extolling Haeckel as one of the greatest Aryan scientists, as "one of the boldest and most essential forerunners of political thought based on natural laws." He commended Haeckel as an example for his fellow Germans: "May the powerful seeker and proclaimer of truth Ernst Haeckel be a pioneer of genuine knowledge, intellectual freedom, and intellectual boldness for the German youth."[19]

When asked by Heberer, Astel enthusiastically agreed to join the Ernst Haeckel Society as a Supporting Member (*Förderer*).[20] When Franz learned that Astel was on board, he immediately wrote to the Nazi Gauleiter of Thuringia, Fritz Sauckel, asking if he would endorse the new society. Sauckel was willing, but first wrote to the Nazi Party leader, Martin Bormann, to make sure this would not be a problem. Bormann replied that he had no reservations about Sauckel supporting the Ernst Haeckel Society, and in September the Nazi ideologist Alfred Rosenberg concurred. Sauckel thus became *Schirmherr* (patron or protector) of the Ernst Haeckel Society with the explicit permission of two of the most influential leaders in the Nazi Party.[21] Among the organizations that joined the new society was the SS Ahnenerbe.[22] The formation of the Ernst Haeckel Society in the 1940s shows conclusively that Haeckel had a good deal of support within the Nazi Party, and he was certainly not persona non grata.

More Honors for Haeckel during the Third Reich

HAECKEL WAS also honored in a 1936 portrait exhibition of "Great Germans" organized by the Reich Minister of Education. This display, set up in a Berlin museum to coincide with the summer Olympic games, included portraits of about 420 "Great Germans." In the book edition of this exhibition, also published in 1936, Haeckel is listed among the great German scientists, and his photo is accompanied by a simple caption: "Zoologist and Philosopher. Teacher at the University of Jena. Enthusiastic proponent of Darwin's theory. Attempted to formulate a popularly accessible worldview on the basis of the scientific knowledge of the nineteenth century (*Riddle of the Universe*, 1899)."[23]

Another photo display of Haeckel in the midst of renowned Germans showed greater ambivalence about Haeckel. In 1941 Karl Richard Ganzer published a book of photos of Germans who represented the leadership of Germany over the past centuries. He included 204 individuals representing a wide ideological spectrum, from religious leaders such as Luther and Schleiermacher, to atheists such as Schopenhauer and Nietzsche. Haeckel's photo was accompanied by a long caption that explained his ideological significance. On the one hand, it praised him for being the leading proponent of Darwinism in nineteenth-century Germany, stating, "It is his greatest achievement, that he, as a courageous fighter who was himself able to point out new ways, advocated for Darwin's revolutionary theory of natural selection: he thereby justifiably challenged an entire generation of scholars." However, after expressing appreciation for Haeckel's contribution to evolutionary theory, Ganzer then criticized Haeckel for "shooting way beyond the target" by trying to erect a new religion based on science. He claimed that Haeckel's monism had been perilously close to materialism and thus was problematic. He concluded, "Haeckel was a bold figure of his time. However, he was unable to plumb the depths and sow eternal values."[24] Not all Nazis would have agreed with Ganzer's critique of Haeckel (as we have already seen), but Ganzer's stance on Haeckel was characteristic of Haeckel's critics: Even they acknowledged and commended him for his contribution to biological evolution. However, his critics found his metaphysical position untenable, because they considered it too close to materialism.

Haeckel in Biology Pedagogy

SCHOOLCHILDREN IN the Third Reich would generally have received a favorable impression of Haeckel in their biology classes. Of the four biology textbooks that the Education Ministry approved for use in German secondary schools in 1939, three mentioned Haeckel, always in a positive light.[25]

In its chapter on "Evolutionary Theory," the Nazi-approved text by O. Steche, E. Stengel, and M. Wagner had a page with illustrations of nine "Great Biologists," and the last three were Lamarck, Darwin, and Haeckel. In the text the authors credited Haeckel for "leading it [Darwin's theory] to victory. Today evolutionary theory is a foundation stone of our worldview."[26] This implies that Darwin and Haeckel were important precursors to National Socialism.

Volume three of Jakob Graf's approved text opens with a section on human evolution, and at the top of this section is an epigram by Haeckel: "The ape-like ancestors of humans have been extinct a long time; perhaps we will find their fossilized bones sometime in Tertiary strata in southern Asia." Thus the very first thing students would see in this book was Haeckel's remarks about human evolution, and the rest of the chapter suggests that Haeckel's views on human evolution had been scientifically confirmed.[27]

Another approved text by Karl Kraepelin, C. Schaeffer, and G. Franke mentioned Haeckel and Darwin for their roles in promoting evolutionary theory.[28] Thus, pupils in Nazi Germany were taught in biology texts that Haeckel was an important biologist for contributing to evolutionary theory.

Biology teachers in the Third Reich frequently met with positive depictions of Haeckel. Ferdinand Rossner, a professor of biology pedagogy at the Hannover Teacher's College and a Nazi Party member, regularly defended Haeckel to his contemporaries. Rossner worked with the Nazi Racial Policy Office in Hannover and played a leading role in the Reich League of Biology, which was connected to the SS Ahnenerbe.[29] In a 1937 book, officially approved by the Education Ministry for use in German schools, Rossner claimed that Haeckel was often misunderstood. He recommended Brücher's laudatory biography of Haeckel and then listed ways Haeckel had contributed to present (i.e., Nazi) ways of thinking: 1) by bringing humans under the sway of nature; 2) by opposing Marxism and equality; 3) by promoting an "aristocratic principle of selection"; and 4) by advocating eugenics.[30]

In 1939 Rossner began editing the *Handbook for Biology Instruction*, essentially an encyclopedia for biology teachers. Two articles in the first volume praised Haeckel for advancing the theory of human evolution. Volume three contained an article on Haeckel by Gerhard Heberer, who began by stating, "He is one of the most important German biologists," and then later explained that Haeckel's *Anthropogenie* (published in English as *The Evolution of Man*) "is one of the great pioneering works of phylogenetic research." Heberer rejected the charge that Haeckel was a materialist and insisted that Haeckel's thought was thoroughly Nordic in character.[31] In a 1942 book Rossner continued to express appreciation for Haeckel for promoting the theory of human evolution.[32]

Biology instructors also found many positive appraisals of Haeckel in *Der Biologe*, the most important German journal for biology teachers. In order to Nazify this journal, in 1934 Hans Schemm, who founded and led the National Socialist Teachers' League, as well as being the Bavarian Minister of Education and a Nazi Gauleiter, was added to the editorial board. Schemm's remark that "National Socialism is politically applied biology" was widely quoted in Nazi Germany. After Schemm joined the editorial board, *Der Biologe* celebrated the hundredth anniversary of Haeckel's birth by devoting an entire issue to his achievements.

Haeckel's great-nephew, Werner Haeckel, contributed an introductory article on "Ernst Haeckel and the Present." Therein he rejoiced that Haeckel's demands for eugenics measures were being taken up by the Nazi regime, since, according to Werner, his great-uncle "demanded the destruction of those with hereditary problems." Werner stated at the close of his brief essay: "May the following essays incite us, with his [Ernst Haeckel's] enthusiasm to apply evolutionary ideas together with the rules of heredity to restore the health of our people, and thus to remember that he brought evolutionary theory, which Darwin founded with his comprehensive works, to a state of approval in the field of biology."[33] The following five articles about Haeckel all praised him for promoting evolutionary theory, including human evolution, and none of them expressed criticism. In one of these articles Friedrich Lipsius took issue with the widespread accusation that Haeckel was a materialist, insisting instead that he was a pantheist along the lines of Goethe.[34]

In 1935 *Der Biologe* became the official organ for the biology section of the National Socialist Teachers' League, and four years later the SS organization Ahnenerbe took over the editorial board, placing Karl Astel, Günther Pancke (head of the SS Race and Settlement Main Office), and other Nazi officials at the helm. After coming under the auspices of the SS, *Der Biologe* commemorated the twentieth anniversary of Haeckel's death in 1939 (two years after Richards claims Haeckel was persona non grata in Nazi Germany) by publishing two articles examining Haeckel's contribution to evolutionary theory, one on zoology and one on botany. Both concurred that Haeckel's phylogenetic trees were essentially correct, including his reconstruction of human phylogeny.[35]

Later that year Rossner published an article in *Der Biologe* on the role of evolutionary theory in biology instruction. He lamented that there were still a few education specialists (not biologists) who were opposed to Haeckel and to evolution altogether. Rossner defended not only evolutionary theory, including human evolution, but also Haeckel by quoting from a positive article published that year in the official Nazi newspaper (see discussion of this below).[36] Also in 1939 *Der Biologe* carried an advertisement for Heinz Brücher's biography of Haeckel that included positive comments about it from *Völkischer Beobachter*, the official Nazi newspaper, and from *Nationalsozialistische Monatshefte*, the Nazi journal edited by Rosenberg. Every mention of Haeckel that I have found in *Der Biologe* was positive, suggesting that he had considerable support among biologists, biology teachers, and at least some Nazi Party officials.

Haeckel's Treatment in Science Journals during the Third Reich

ANOTHER JOURNAL that honored Haeckel during the Nazi period was the *Archiv für Rassen- und Gesellschafts-Biologie*, edited by the prominent eugenicist Alfred Ploetz. For many years it was the official organ of the German Society for Racial Hygiene, and just a few months after the Nazis seized power, it was put under the authority of the Reich Committee for People's Health Service, an organization within the Nazi-controlled Ministry of the Interior. Ploetz admired Haeckel considerably, and when he had originally organized the Society for Racial Hygiene in 1905, he recruited Haeckel as an honorary member. In the fifth issue of 1936 Ploetz published a full-page photo and inscription of Haeckel on the opening page of the journal. The following year Ploetz's journal published Franz's inaugural lecture on Haeckel's evolutionary theory. Franz called himself an "enthusiastic admirer of Haeckel" and urged his fellow scientists to build on Haeckel's work.[37]

In his zeal to distance Haeckel from Nazism, Richards highlights two 1937 articles from the *Zeitschrift für die gesamte Naturwissenschaft*, a journal published under the auspices of the science section of the National Socialist Student League. The most important of these was by Günther Hecht, and according to Richards, "The efforts to recruit the author of *Die Welträthsel* [i.e., Haeckel] to the Nazi cause foundered almost immediately because of a quasi-official monitum issued by Gün-ther Hecht, who represented the National Socialist Party's Department of Race Politics (Rassenpolitischen Amt der NSDAP)."[38] I have no idea why Richards thinks an article in a science journal by a lower-level Nazi functionary constitutes a "quasi-official monitum." Nonetheless, Rich-ards is correct that Hecht strongly criticized Haeckel as a purveyor of materialistic ideology and warned his fellow Nazis against Haeckel's "materialistic monism."

However, Richards fails to mention that in his essay Hecht also ac-knowledged the positive contributions Haeckel made to science. While dismissing natural selection in Darwin's and Haeckel's evolutionary theory as too mechanistic, Hecht nonetheless believed in evolution, stat-ing, "The fundamental ideas of *evolutionary theory*—according to which all living organisms on our earth descend from ancestors and forebears, and in the course of geological history gradually develop from older and simpler forms—rest on what is today an altogether uncontested real-ity." Indeed, even though he thought Darwin's and Haeckel's reliance on natural selection was misguided, he still honored both men for promot-ing evolutionary theory: "Viewed from today's standpoint in relation to worldview and science, the age of *Darwin* and *Haeckel* in biology was es-sentially the age of the final triumph of the (correct) theory of evolution and descent, to which both scientists contributed decisively." Hecht also clearly believed in human evolution, though he suggested that further research was needed to determine the human phylogenetic tree.[39]

Interestingly, despite his criticisms of Haeckel's monism, Hecht was not criticizing monism per se. He was only opposing what he perceived as the mechanistic and materialistic thrust of Haeckel's form of mo-nism. In the opening paragraphs of his essay, he roundly rejected body-soul dualism and criticized Pauline Christianity for embracing human equality. He implicitly advocated a form of monism, but an idealistic, anti-materialistic form.

The article in the *Zeitschrift für die gesamte Naturwissenschaft* by University of Kiel philosophy professor Kurt Hildebrandt earlier in 1937 raised the same objections against Haeckel. Like Hecht, Hildebrandt explained that he believed in biological evolution, including human evolution, but he thought Haeckel's theory of evolution was too mechanistic and thus materialistic. Hildebrandt suggested that a pantheistic worldview was the proper antidote to Haeckel's materialism. Thus, Hildebrandt was not rejecting monism, but rather was objecting to what he saw as Haeckel's materialistic version of monism. For both Hecht and Hildebrandt, the problem with Haeckel was not that he promoted evolution or monism, but that he was—at least in their perception—a philosophical materialist.[40]

Despite Richards's claim, Hecht's essay was not the final, official Nazi position on Haeckel. Indeed Otto Reche, an anthropology professor at the University of Leipzig who frequently lectured on racial science to Nazi Party organizations, was so outraged by Hecht's article that in December 1937 he wrote to Walter Gross, the leader of the Racial Policy Office (thus Hecht's boss), to protest it. That same day Reche also wrote to Heberer, stating, "One needs to try to hinder further idiocies from Herr Hecht; it would be best if he would resign from the staff of the Racial Policy Office."[41] Neither Reche nor many other Nazi Party members and officials considered Hecht's article a final pronouncement. Rather it was just one salvo in an ongoing debate over Haeckel in Nazi Germany.

Unfortunately, I have yet to locate a record showing whether or how Gross responded to Reche's harsh critique of Hecht. However, in other utterances, Gross evinced some sympathy with Haeckel, though in a cautious way. In 1934 Gross told a National Socialist Teachers' League meeting in Leipzig:

> Many teachers go down intellectual false paths. They believe they can solve all the questions of life with Haeckel's *Riddle of the Universe* and the *Monistische Monatshefte*. If they really believe that, then they are

really very foolish. From the foreword to the *Riddle of the Universe* book they could have gathered how little Haeckel was claiming to have solved all the riddles of the universe. And in a similar way, as far as I know, the German Monist League always explained this, too. It is our conviction, of course, that with Haeckel's monism we come closer to a solution of the riddle of the universe than with other ways of thinking.[42]

Here Gross was clearly trying to put a damper on Haeckel worship, and he seems to have wanted to put some distance between Haeckel and Nazism. However, in the final sentence of the quotation he admits that he is not an outright foe of Haeckel's monism, but rather considers it congenial to his worldview.

Around 1939 in a private memo Gross also expressed caution about Haeckel's views being identified with Nazism. At that time Gross was wrestling with how far he should go in cooperating with Hans Weinert, an evolutionary anthropologist (and SS officer) whom Gross considered "the most competent specialist in the field of the theory of human origins," according to Ute Deichmann. Gross wrote that Weinert had never

directly presented his theory of evolution as part of National Socialist ideas on race. On the other hand, it is my opinion that the political racial doctrine of National Socialism should maintain sufficient distance from the doctrine of human evolution, which... still subject to the changes of scientific theories, is frequently still pervaded with Haeckelian ways of thinking in its basic ideological ideas... and is thus publicly considered a part of materialistic, monist ideas.[43]

Gross was obviously not rejecting human evolution altogether here, since he was generally favorable toward Weinert's evolutionary anthropology (and a few years later Gross published a scholarly article devoted entirely to human evolution).[44] Gross's primary concern with Haeckel seemed to be the public perception that Haeckel was a materialist, and thus he did not want Nazi racial theory to be tainted by this association.

Anthropologists' Views of Haeckel

SOME LEADING anthropologists publicly praised Haeckel throughout the Nazi period. Heberer, who helped found the Ernst Haeckel Society, lectured and wrote incessantly about Haeckel, whom he esteemed greatly. While a lecturer at the University of Tübingen in 1934, he honored Haeckel by publishing a lecture on his hundredth birthday. He noted that modern anthropology was based on Haeckel's evolutionary insights, and he argued that all of Haeckel's scientific work was essentially correct. He also reminded his audience that Haeckel had been "one of the first proponents of eugenics measures," which the current regime was implementing.[45] That same year Heberer also taught an entire course on Haeckel at the University of Tübingen.[46] In 1937–38 Heberer wrote two articles about human evolution in *Volk und Rasse*, an anthropological journal. In both he confirmed that he agreed with Haeckel's views on human evolution and defended Haeckel against those who thought his ideas were materialistic. Heberer insisted that Haeckel's critics did not understand Haeckel correctly.[47] Heberer also wrote several articles and reviews for *Der Biologe* about Haeckel.

Other anthropology professors who joined the Ernst Haeckel Society in the 1940s included Reche (University of Leipzig), Weinert (University of Kiel), and Theodor Mollison (University of Munich).[48] In 1944 Weinert published the second edition of his *Ursprung der Menschheit* (*Origin of Humanity*), which was a detailed study of human evolution. While he did not endorse all the details of Haeckel's scientific theories, he stated that Haeckel's ideas were fundamentally still valid. Not only did he defend Haeckel's view that humans descended from ape-like ancestors, he even dedicated the volume to Haeckel.[49] Reche had studied under Haeckel for one semester at the University of Jena in 1902 and was heavily influenced by Haeckel, so it is not surprising that he was a leading member of the Ernst Haeckel Society.[50]

Hans F. K. Günther, the racial anthropologist with the closest connections to the Nazi regime, was more conflicted about Haeckel and refused to join the Ernst Haeckel Society. When Franz asked him to join, Günther replied that he had read Haeckel's *Riddle of the Universe* as a student, but he found it philosophically shallow, even though he agreed with Haeckel's opposition to Christianity. Thus he did not want to be too closely associated with Haeckel personally. Nonetheless, he was willing for his university institute to join the Ernst Haeckel Society, so he was not completely opposed to Haeckel.[51]

While privately distancing himself from Haeckel's philosophy, on occasion Günther publicly mentioned Haeckel in a more favorable light. In the opening paragraph of his inaugural lecture at the University of Berlin in 1935 he credited Haeckel with having understood the aristocratic implications of evolutionary theory. Thus he connected Haeckel with the Nazi leadership principle (*Führerprinzip*), arguing that Haeckel's evolutionary theory contributed to the notion of a "leadership aristocracy" (*Führeradel*).[52] In another lecture from 1935 Günther explicitly connected Haeckel's ideas to Hitler. He quoted a statement in *Mein Kampf* where Hitler praised the "aristocratic foundational principle of nature." Günther traced this idea of nature as aristocratic back to Haeckel. He also mentioned Otto Ammon and Alexander Tille as important precursors to Hitler's idea,[53] but Ammon and Tille were themselves heavily influenced by Haeckel's attempts to apply Darwinism to society.[54]

In a 1936 lecture Günther commented that democrats and socialists were wrong to try to appropriate Haeckel, because Haeckel had supported aristocratic conclusions from his evolutionary theory.[55] Based on this and other of Günther's writings, it is clear that while Günther thought Haeckel's attempts at being a philosopher had miscarried, he agreed with Haeckel on many points, including evolutionary theory and at least some elements of his social Darwinism.

Haeckel in Nazi Periodicals and Propaganda

NATIONAL SOCIALIST periodicals occasionally carried articles on Haeckel, and in most cases, they expressed esteem and appreciation for him. As we saw in Chapter 6, in 1934 and 1939 the official Nazi newspaper *Völkischer Beobachter* carried articles lauding Haeckel.[56] The latter one enthusiastically celebrated Haeckel's life and work, especially his contribution to the theory of human evolution. The author insisted that those critics who portrayed Haeckel as an atheist or materialist were distorting Haeckel's views. Then the author exulted that Haeckel was being rehabilitated, suggesting that his image was improving in the late 1930s in Nazi Germany. The author also argued that Haeckel was an important precursor to Nazism: "After the argument has been fought out over the details of his work that are time-dependent, the figure of this man appears to us as historically significant in gaining acceptance for his biologically grounded worldview and thus also for the intellectual pre-history of the National Socialist revolution." According to the *Völkischer Beobachter* in 1939, Haeckel's star was still rising.[57]

Alfred Rosenberg's *Nationalsozialistische Monatshefte* published an article about Haeckel by Brücher in 1935. As indicated in the title, "Ernst Haeckel, a Forerunner of Biological Political Thought," the article hailed Haeckel as an important formulator of ideas that would come to fruition in Nazi ideology. Specifically, he credited Haeckel with laying the basis for current (i.e., Nazi) racial theory, as well as promoting eugenics, including killing those deemed unfit. He also praised Haeckel for rejecting human equality, body-soul dualism, and left-wing socialism. According to Brücher, "As the first thinker of the Western world Ernst Haeckel drew wide-reaching worldview conclusions from the scientific knowledge of the origin of humans and their evolution according to the laws of the "selection of the best" and the "survival of the fittest"; this led him to reject that politically dominant idea of the equality of humans and their plasticity through education."[58] Many of Brücher's arguments in this essay were also reflected in his biography of Haeckel that came out the following year. Apparently Brücher's article aroused some opposition from the Catholic press, because in 1936 *Nationalsozialistische Monatshefte* carried another article by Brücher defending Haeckel from these Catholic critics.[59]

The official periodical of the Hitler Youth, *Wille und Macht*, edited by Baldur von Schirach, also carried a 1936 article that portrayed

Haeckel as an important forerunner of Nazi ideology. This article by Ernst Lange focused on the racial ideologue Ludwig Woltmann, who constructed a social Darwinist ideology that powerfully impacted racial thought in early twentieth-century Germany, influencing anthropologists such as Eugen Fischer and Otto Reche (the latter republished Woltmann's social Darwinist works during the Nazi period). Lange mentioned the impact of Darwin and Haeckel on Woltmann and then claimed, "Today both scientists [Darwin and Haeckel] are indisputably regarded as the fathers of modern biological thought, racial science is built on their science, and these great scientists only celebrate their true victory today, since an entire Reich has begun to set up its highest principles according to the results of biological research."[60] Lange's article suggested that Darwin, Haeckel, and Woltmann were all important intellectual precursors to Nazi ideology.

One of the most stunning anti-Haeckel pronouncements in Nazi Germany came in 1935 from a Saxon state agency in charge of libraries. In the official journal for librarians, Die Bücherei, this agency published a list of categories of books that should be banned from libraries. The sixth blackballed category was "Writings of a worldview or biological character, whose content is the superficial scientific enlightenment of a primitive Darwinism and monism (Haeckel and his followers, also obsolete Ostwald)."[61] This is an important example of opposition to Haeckel in Nazi Germany, but Richards goes far beyond the evidence when he claims, "The rejection of Haeckelian ideas had been sealed in 1935 when the Saxon ministries of libraries and bookstores banned all material inappropriate for 'National-Socialist formation and education in the Third Reich.'"[62] As I have already shown and will continue to show, the debate over Haeckel was by no means over in 1935, nor in 1937. He had many prominent Nazi defenders, and various indications suggest that his reputation was actually improving after 1937.

Also, it is not clear how influential this proclamation by a Saxon state library official was. During the Nazi period several different lists of banned books were drawn up by Nazi officials, and the others did not include Haeckel's works.[63] Further, when Franz published the first volume of the yearbook of the Ernst Haeckel Society, he listed the books by and about Haeckel that were still in print in 1943. This list included twenty-nine books by Haeckel published by five different publishers, including his two most important works, *Natürliche Schöpfungsgeschichte* (*Natural History of Creation*, published by Walter de Gruyter in Berlin) and *Welträtsel* (*Riddle of the Universe*, published by Alfred Kröner in Stuttgart). Brücher's biography of Haeckel (published by J. F. Lehmann in Munich), which portrayed Haeckel as a forerunner of Nazism, was also still in print in 1943 (and it was initially published in 1936, thus after the Saxon library official published the blacklist).[64] It is unknown if or how many libraries may have removed copies of Haeckel's works, but even if some did, Haeckel's books were still being sold in Nazi Germany.

Haeckel's staunchest and most influential critic in Nazi Germany was Ernst Krieck, who joined the Nazi Party in 1932 and became professor of philosophy and pedagogy at the University of Heidelberg. Krieck was an implacable foe of materialism, and like many of Haeckel's critics, he interpreted Haeckel as a materialist. However, like many others we have examined, Krieck upheld a monistic metaphysics and explicitly rejected all varieties of metaphysical dualism. He spurned the mechanistic vision of the universe, instead embracing a pantheistic view of the cosmos as a living organism.

Among Haeckel's Nazi critics, Krieck was idiosyncratic by challenging the biological theory of evolution altogether. Krieck objected to the way that Darwin's theory removed teleology and purpose from the universe, which he considered essential characteristics of the living cosmos. However, Krieck's criticisms of biological evolution were not intended to buttress a Christian or creationist perspective. Indeed, he resoundingly rejected what he called the "oriental myths" of the Old and New Testaments. He insisted that neither science nor religion could provide any answers about origins, so Krieck expressed agnosticism about biological origins.[65] Krieck's criticisms of Haeckel were utterly rejected by Alfred Rosenberg and his office.[66] Also, according to Hans Sluga, on the whole Krieck's influence waned during the Third Reich.[67]

Conclusion

THIS EXAMINATION of the Nazi debates over Haeckel shows a wide divergence of views. While several saw Haeckel as an important precursor of Nazi ideology, a few others depicted him as a dangerous opponent of Nazism, so dangerous that his books should be banned from libraries. Those who saw Haeckel as a forerunner of Nazism focused on several key ideas of Haeckel's that foreshadowed Nazi ideology and policies: 1) biological racism; 2) eugenics; 3) euthanasia; and 4) rejection of socialism and human equality. Haeckel's detractors consistently accused him of materialism, while his supporters denied that he promoted materialism.

Interestingly, both his critics and fans agreed on several points. First, they all agreed that philosophical materialism was taboo in Nazi thought. No one argued that it was acceptable for Haeckel to be a materialist. However, most of Haeckel's Nazi critics who explained their own metaphysics tended toward some kind of non-materialistic monism, such as pantheism, so they were not anti-monistic. Second, except for Krieck, almost everyone debating about Haeckel accepted biological evolution, including human evolution. Many Nazis praised Haeckel for his contributions to evolutionary theory, and almost none criticized him for teaching about evolution. Finally, none of his Nazi detractors criticized Haeckel for being hostile to Christianity. (Haeckel did have some opponents in the Catholic press, but my focus here is on the Nazi perceptions of Haeckel, not the debate over Haeckel in non-Nazi venues).

In sum, Haeckel was not the most important progenitor of Nazi ideology, and not even his most ardent Nazi supporters portrayed him as such. He was certainly not the source for the vitriolic anti-Semitism that played such a key role in Nazi ideology.[68] However, he did contribute to some features of Nazi ideology, such as biological racism, eugenics, and killing people with disabilities. Haeckel was also by no means persona non grata in the Third Reich, despite opposition from a few lower-level Nazi officials. High-ranking officials, such as Rosenberg and Bernhard Rust, were clearly sympathetic with Haeckel, and neither Rosenberg nor Bormann saw any problem with the formation of the Ernst Haeckel Society in 1941. The Nazi press, Nazi school texts, and science periodicals were far more positive than negative in their portrayals of Haeckel. Though Haeckel was a disputed figure in the Third Reich, it seems that on the whole he was admired more than denigrated, and his evolutionary views championed.

8. Darwinism in Neo-Nazism and White Nationalism

THE COLUMBINE HIGH SCHOOL MASSACRE, DESCRIBED IN THE INtroduction, was not the only white nationalist mass shooting motivated by Darwinian impulses. On July 28, 2019, a nineteen-year-old gunman opened fire at the Garlic Festival in Gilroy, California, killing three people and wounding seventeen. After being shot by police, he turned his gun on himself. Shortly before perpetrating his evil deed, he posted to his newly established Instagram account a suggestion: "Read Might Is Right by Ragnar Redbeard." He then followed this up with a racist comment. The subtitle to *Might Is Right* is *Survival of the Fittest*— the nickname for natural selection coined by Herbert Spencer and adopted by Darwin himself as a description of his theory. Redbeard's book is the most incendiary piece of social Darwinist literature that I have ever read, which is saying a lot, because I am an expert on the history of social Darwinism. Perhaps that is why the author used a pseudonym.

Redbeard's book appeared in 1896, but it does not seem to have been all that influential in the 1890s or early decades of the twentieth century. Interestingly, it began circulating more widely after 1969, when Anton Lavey, founder of the Satanic Church, plagiarized portions of it in his *Satanic Bible*. Lavey, who later wrote a preface for an edition of *Might Is Right*, dedicated early editions of the *Satanic Bible* to Redbeard. He claimed that when he discovered this book in a used bookstore in 1957, it was pretty much unknown and would have remained obscure if he had not publicized it.[1]

Might Is Right has become very popular among white nationalists. The website Aryan Resistance has a free pdf version of the book on its website, and many other white nationalist websites either sell the book or recommend it. There are a couple of websites devoted exclusively to Redbeard's ideology. In September 2009 a retired psychology professor, Andrew Hilton, published a positive review of Redbeard's book at the Occidental Observer, a prominent white nationalist online journal. Hilton not only calls Redbeard well-informed and erudite, but he also overtly endorses Redbeard's social Darwinism. Hilton embraces evolutionary psychology, the doctrine that many human behaviors are hereditary and can be explained as an adaptation in the Darwinian struggle for existence. He dismisses universal morality as maladaptive in the evolutionary process.[2]

On the Storm Front website, an important white nationalist discussion forum, James Harting posted a blog in September 2019 on Redbeard's book, encouraging his audience to read it (he also recommended Nietzsche's *Anti-Christ* and the Nazi ideologist Alfred Rosenberg's *Myth of the Twentieth Century*). Harting noted, "As I have written elsewhere on Stormfront, this book [*Might Is Right*] played an important role in the formation of my own worldview many years back."[3]

In the pages of *Might Is Right*, Redbeard encouraged readers to grasp for power in order to triumph in the merciless Darwinian struggle for existence among unequal humans. From the very first pages Redbeard viciously and relentlessly attacked Christianity—calling Jesus "the true prince of Evil"—and spurned Christian morality, such as loving others, showing compassion, and helping the weak. Rather, Redbeard insisted, "In actual operation Nature is cruel and merciless to men, as to all other beings.... We must be, like nature, hard, cruel, relentless." He dismissed with contempt the US Declaration of Independence's dictum that "all men are created equal, that they are endowed by their Creator with unalienable rights," because, he claimed, the Creator it referred to is a "mythical airy being." Rather than equality, he portrayed some humans—such as black Africans, East Asians, and Jews—as inferior to the "Aryan" race.[4]

Right after mentioning that races are not part of some universal brotherhood, Redbeard stated, "And it is proposed to prove in this book, that strife, competition, rivalry, and the wholesale destruction of feeble types of men is not only natural, but highly necessary." These allegedly unequal humans are locked in mortal combat, and "instinctively we understand that the struggle for existence is absolutely needful."[5] Not only that, but—as the title of the book also indicates—the only yardstick for morality is whatever helps one win the struggle for existence. Redbeard explained, "When not thwarted by artificial contrivances, whatever argument Nature promulgates is—*right*.... To be right is to be natural, and to be natural is to be right.... Darwin's law exists—may be seen in operation—is practicable—of daily demonstration—therefore it also is right."[6]

Redbeard reveled in fighting, competition, and strife, because he thought it would produce further evolutionary progress.

Death was a good thing, he insisted, because it would rid the world of the allegedly inferior biological specimens. He stated, "Cursed are the Unfit for they shall be righteously exterminated." He considered this beneficial, because it would promote further evolution. He wrote, "Superiority can only be decided by Battle. Conflict is an infallible method of Selection and Rejection. Evolution has no end. That is undoubtedly, the logical deduction of Darwin's famous pronouncement: 'If he (man) is to advance still higher, it is to be feared that he must remain subject to a severe struggle. Otherwise he would sink into indolence; and the more gifted men would not be more successful than the less gifted.'"

That quotation from Darwin's *Descent of Man* was just one of many quotations and explicit allusions to Darwin. Indeed, Redbeard once enjoined his readers: "Be a Darwin in active operation."[7] It should not be surprising that Redbeard's radical social Darwinism resonates with many white nationalists, particularly the many among them who see

136 / <small>Darwinian Racism</small> /

Hitler as a positive role model. After all, Hitler embraced, and acted upon, many of the same ideas. Indeed, some of Hitler's followers in the United States preached the Darwinian message of racial struggle even before Redbeard's book had gained currency among white nationalists.

George Lincoln Rockwell's American Nazi Party

Biological evolution played a central role in the worldview of George Lincoln Rockwell, who founded the American Nazi Party in 1959. He led the party until he was assassinated by a disaffected former member in 1967. A year before he died he published *White Power*, a compendium of his racist worldview, which is still available on the party's website.

Rockwell—like his role model Hitler—seems to deify nature, telling his followers that they should cultivate a "worshipful attitude toward Nature." He consistently capitalized Nature in his book, and he claimed that our highest goals are to follow the ways of Nature. He stated, "Our all-out belief in race, our insistence on the natural laws in society, economics and every other field of human activity are, in every case, the conscious, scientific application of Nature's iron laws, instead of conceited and short-sighted perversions of these laws, as pushed by the arrogant, peanut-brained liberals and Marxists."[8]

One of the natural laws that Rockwell emphasized repeatedly was natural selection through the struggle for existence, which brings about human evolution. He stated, "Nature took care that humans keep evolving, by eliminating the unfit and breeding the race ever upward, in spite of human egotism."[9] Of course, natural selection involves death to the unfit, and Rockwell did not recoil from this cruel side of nature. When discussing the Malthusian population imbalance (on which Darwin founded his theory of natural selection), Rockwell coolly explained, "There is no 'cure' for the coming population horror other than to kill. Nature did the killing, by natural selection, since the beginning of time. The birth-control advocates, with typical liberal cowardice and short-sightedness, do their killing by un-natural selection, by cowardly murdering before the people they kill have any chance to argue the case, or prove themselves."[10] So who does Rockwell think should do the killing and who does he think should be killed? Well, there is a reason the book is entitled *White Power.*

Rockwell stated that racial disparities arose through biological variation in the evolutionary process. His explanation of how races diverged is remarkably (but not surprisingly) similar to the story told by Nordic racists (and Hitler himself) in the early twentieth century. According to Rockwell, "A natural selection of men occurred when the energetic ones left the warm climates where man originated, leaving the lazy ones behind. In the North, man had to think ahead to live. The foresighted and unselfish people of the north then bred with each other to produce still more foresighted, resourceful and unselfish people."[11] Rockwell, like most Nordic racists, believed that the evolutionary selective pressures on people living further north caused them to evolve a higher level of intelligence, as well as a higher morality. The continuing upward evolution of the white race—which was one of Rockwell's central goals—was imperiled, he thought, by the ideological trends promoted by liberals and leftists. He explained, "But if the race of Negroes is, as a matter of FACT, INFERIOR, then it is the other way around and the liberals, Jews, Communists and egalitarians are the wicked poisoners of a million years of White evolution and breeding."[12] The evolution of human races through natural selection thus plays a central role in Rockwell's neo-Nazism.

The New Order, one of the successor organizations to the American Nazi Party, continues the legacy of Rockwell. On its website it answers the question, "What is National Socialism?" by enunciating seven principles that Rockwell had developed. The first point explains that the "goal, for us, is the upward struggle of our [white] race and the fight for the common good of our peoples." The third principle states, "We believe that man makes genuine progress only when he approaches Nature humbly, and accepts and applies her eternal laws." The next principle clarifies that the Darwinian struggle for existence is one of the most important laws of nature: "We believe that struggle is the vital element of all evolutionary progress and the very essence of life itself."[13]

On another of their webpages that provides an introduction to the New Order, it articulates "three idealistic principles": nature, race, and personality. The first of these is that nature is the ultimate authority. The second one, the racial principle, is: "We believe in the ultimate sanctity of White (or Aryan) blood, and in the necessity for upward evolution of life toward a higher state."[14] Thus, Rockwell's social Darwinist principles are still being promoted zealously by his followers.

William Pierce and the National Alliance

Another white nationalist leader who placed Darwinian evolution at the center of his worldview was William Pierce. From the 1970s until his death in 2002 he was probably the most influential white nationalist in the United States. In a book analyzing white nationalism, the religious studies professor Damon Berry argues that "there is really no way to overstate Pierce's importance to contemporary white nationalism."[15] In the 1960s Pierce participated in Rockwell's American Nazi Party, editing the party's journal, but after Rockwell's death in 1967, he left the party. In the early 1970s he joined the National Youth Alliance, and in 1974 he reorganized it as the National Alliance, which he led until his death. Pierce also founded a new religion, Cosmotheism, which promoted the veneration of nature. In a 1977 speech he admitted that his new religion was pantheistic, but it differed from the views of many earlier pantheists, who "lacked an understanding of the universe as an *evolving* entity and so their understanding was incomplete."[16] In that speech he also explicitly praised Darwin as the key thinker shaping his view of nature.

Today the National Alliance continues to emphasize the importance of evolutionary theory in shaping its goals and policy proposals. Though this organization declined in influence after Pierce's death in 2002, it reorganized in 2013–14 and regained some impetus. On its webpage entitled "What Is the National Alliance?" they begin by stressing the supremacy of nature.[17] The first section is on "The Natural World," which begins by stating, "We see ourselves as integral with a unitary world around us, which evolves according to natural law. In the simplest words: There is only one reality, which we call Nature." The second section is on "The Law of Inequality," which teaches that each human race is different, because it has "developed its special characteristics over many thousands of years during which natural selection not only adapted it to its environment but also advanced it along its evolutionary path." Those races having to confront harsher northern climates allegedly "advanced more rapidly in the development of the higher mental faculties—including the abilities to conceptualize, to solve problems, to plan for the future, and to postpone gratification"—than people remaining in tropical climates. "Consequently," we are told, "the races vary today in their capabilities to build and to sustain a civilized society and, more generally, in their abilities to lend a conscious hand to Nature in the task of evolution."

The National Alliance hopes to promote further evolution by advancing the interests of the white race and achieving a "White Living Space." (This term "living space" was coined by the Darwinian biologist-turned-geographer Friedrich Ratzel in the late nineteenth century, because Ratzel thought the essence of the Darwinian struggle for existence was a "struggle for space.")[18]

Today the *National Vanguard*, the official journal of the National Alliance, explains succinctly on its website what it stands for. The opening words declare:

> Science continues to prove that race exists. The Human Genome Project has unlocked human DNA, and medical advances are being made based on targeting genetic sequences individual to each race. Thousands upon thousands of years of separate racial development have led to a great diversity of human races, which have in turn formed separate cultures with vastly different values and achievements. *Race, variety,* and *subspecies* are all similar terms and they apply to human beings no less than to other life forms on this planet. Races are new species in the making. *Without racial divergence, the evolution of life itself could never have taken place.*[19]

Of course, when they say racial differences, they do not mean separate but equal, but rather white supremacy. They clarify, "The European race is uniquely beautiful and creative. It is imperative that we survive and progress."[20]

In July 2018 one of the associate editors of *National Vanguard*, Kevin Alfred Strom, who was a disciple of Pierce and is a leading figure in the National Alliance, published "Our Evolutionary Morality." Strom regards evolution as an unshakable truth essential to our understanding of humanity. Not only that, he believes that our morality has to be based on evolutionary principles. He explains:

> If you want to be an agent to restore true morality to our people, you must understand certain things. The first is the evolutionary nature of life. Life is ever-evolving, ever-changing. Evolution is real. Evolution is not only real, it is a fundamental, necessary, aspect of life—in fact, without evolution, living things as we know them could not exist. And racial divergence and separation are themselves *essential* parts of evolution. *Without racial separation, therefore, there could be no life at all.*[21]

From here Strom concludes that racial separation is a command imposed on us by nature. "Since racial separation is essential to life, I believe it is proper to say that, if anything is sacred, racial separation is sacred," he claims. The reason he thinks racial separation is sacrosanct is that he believes nature wants us to promote evolutionary progress. According to Strom, his Cosmotheist ideology "regards our survival and upward evolution as imperatives."[22]

Strom's article is not at all atypical for the magazine. In addition to its primary thrust, which is white supremacy and biological racism, the *National Vanguard* also carries many articles promoting biological evolution, evolutionary ethics, and eugenics. In September 2019 it published a portion of Richard McCulloch's white nationalist book, *The Racial Compact*. This excerpt carried the title "The Evolutionary Basis of Racial Diversity." McCulloch taught that geographic isolation was the driving force in biological evolution, producing racial diversity, and warned that racial interbreeding would hinder "the creative process of evolutionary divergence."[23] In 2015 James Hart's article "God, Evolutionary Ethics, and Eugenics" complained that currently humanity was devolving, rather than evolving. He hoped that humanity could be rejuvenated through evolutionary ethics and eugenics, because "good is what improves us and evil is what weakens or destroys us."[24]

National Vanguard has published quite a few articles by H. Millard that discuss human and racial evolution. In "They Want to Stop Our Evolution," he urges his fellow white nationalists to promote the further evolution of the white race by separating from other races, who he claims are trying to halt evolutionary progress by miscegenation.[25] In an earlier article Millard claimed that those who truly understand evolution and the laws of nature will recognize that Hitler and the Nazis were not evil at all. Indeed, he argues, "Whites would be much better off today had the National Socialists won [World War II]."[26]

Another editor at *National Vanguard*, Mary W. Pennington, has cross-posted articles about Darwinism and racial evolution from other websites, including an article from the "Human Evolution News" website, which promotes "subspecieism."[27] According to that website, "Subspecieists believe that there is vast diversity among modern humans and that our differences shouldn't be downplayed or concealed." This website rather disingenuously claims that its perspective is not racist, but rather celebrates racial diversity. Though it is not overtly promoting white supremacy, as other white nationalist websites do, it emphasizes many of the same themes as white nationalists, including homage to Darwinism: "First and foremost, subspeciests [sic] are committed Darwinists."[28]

In the past few years the *National Vanguard* also has reprinted articles by an extremely influential figure in the twentieth-century white nationalist movement, Revilo P. Oliver, a professor of classics at the University of Illinois until retiring in 1977. Damon Berry calls Oliver "one of the most important figures in the development of American white nationalism."[29] Oliver, who was a friend of Pierce, wrote many books and articles promoting white racism, while bashing Jews and Christians. In 2017 *National Vanguard* republished a 1986 piece by Oliver, "To Honor Darwin," that congratulated the Smithsonian Institution for issuing a book on evolution, which—according to Oliver—delivered a "handsome rebuke to the Jesus-jerks." He opened the article by stating: "THE THEORY of biological evolution requires no proof. It is the only reasonable and logically plausible explanation of the origin and development of life on this planet that has been thus far proposed. For rational men, there is no alternative." The Smithsonian failed, however, in Oliver's view, to point out "the great physical and greater mental and spiritual differences between extant races."[30] In another article on "The Piltdown Forgery" Oliver also lambasted creationism.[31]

Pierce's Cosmotheist Church continues to exist, too. Their webpage entitled "What Is Cosmotheism?" provides a pithy description of their beliefs: "Cosmotheism is the religion of upward evolution. Cosmotheism is the religion of European Civilization. Cosmotheism is the religion of reality and science." The Cosmotheist website sells not only books by Pierce, but also Hitler's *Mein Kampf* and Ragnar Redbeard's *Might Is Right*. Unsurprisingly, Cosmotheism continues to propagate the social Darwinist vision of both Hitler and Pierce.[32]

Ben Klassen and His Darwinian Racial Holy War

IN THE late twentieth century Ben Klassen called on his fellow white nationalists to engage in a Racial Holy War, often shortened to RaHoWa by proponents. In the 1930s he was heavily influenced by reading *Mein Kampf*, and Hitler's ideas about racial struggle are reflected in his own writings. Klassen organized his white nationalist disciples by founding a new religion in 1973, the Church of the Creator (renamed the World Church of the Creator after Klassen's suicide in 1993).[33] Klassen's followers did not always interpret the Racial Holy War in a metaphorical sense, for several of them have engaged in acts of violence against non-whites, both before and after Klassen's death. In 1991, for instance, George Loeb, a leader in the Church of the Creator, shot to death a black sailor in Florida. Other members of the Church of the Creator have been arrested for bomb plots, shooting sprees, and other acts of violence.

Klassen's religion, which he dubbed Creativity, was pantheistic and resembled Pierce's Cosmotheism. The same year he founded the Church of the Creator, he wrote *Nature's Eternal Religion* to explain his religious creed. The Creator, as it turns out, is Nature itself, which Klassen always capitalized in his writings. He proclaimed that the White Race (also capitalized in this work) is "the pinnacle of her creation." The central focus of Creativity is on the laws of nature, which are immutable and must be followed. Indeed, "one of Nature's most inexorable laws is the survival of the Fittest." In Klassen's view, humans cannot opt out of this pitiless struggle, but should emulate nature. He asserted, "In the struggle for the survival of the species Nature shows that she is completely devoid of any compassion, morality, or sense of fair play, as far as any other species is concerned. The only yard stick is survival." Thus Klassen was promoting a kind of evolutionary ethics, where whatever promotes survival and reproduction is deemed good, and whatever is weak or sickly is evil.[34]

According to Klassen, Nature favors the "inner segregation of species," a phrase he borrowed from Hitler's *Mein Kampf*. However, just like Hitler, he erroneously applied this principle to races, as well as species. Thus, he promoted racial segregation and opposed miscegenation. He wanted white people to reproduce as abundantly as possible, stating, "The very struggle for existence for any species, and at this stage in history, particularly the White Race, depends on how prolific and how fruitful each and every one of us is in bringing in the next generation." He adjured his fellow white nationalists to help "propel humanity to an ever-higher level of evolution. Only the White Race can do this."[35]

Evolutionary Psychologists Promoting White Nationalism

Two of the most influential academics to support white nationalism in the 1990s and thereafter were the psychology professors J. Philippe Rushton and Kevin MacDonald. Rushton's 1995 book, *Race, Evolution*

and Behavior, has exerted a powerful influence on the white nationalist—or alt-right—movements (more on the alt-right below). Jared Taylor, who is himself a leading writer in alt-right circles, published a review
of Rushton's book in December 1994. He began by exulting, "*Race, Evolution, and Behavior* is one of the most important books about race to be
written in many years."[36] Later, in 2017, Taylor was still recommending
Rushton's book as the most important non-fiction book for white nationalists to read.[37] In 2014 another influential writer in the white nationalist movement, F. Roger Devlin, placed Rushton's work at the top
of his recommended reading list.[38]

Rushton, who was a professor of psychology at the University of
Western Ontario, attempted to use evolutionary psychology to explain
racial differences. For purposes of his analysis, Rushton divided humanity into three races: whites, blacks, and east Asians. Based on intelligence
testing, he argued that east Asians are the most intelligent, whites are
close behind, and blacks are far less intelligent. (It should be noted that
intelligence test scores have varied significantly over time even within
racial groups, that racial disparities have not remained static, and that
blacks raised by white parents have on the average much higher scores
than blacks raised by black parents, strongly suggesting that the test
score differences are driven by cultural/sociological factors rather than
biological ones. For these and other reasons, the vast majority of psychologists reject Rushton's misguided claims.)[39]

Rushton also claimed that east Asians and whites are more cooperative and moral, while blacks have greater inclinations to crime and im-
morality. He, like many white nationalists, also stressed the physical differences between races, especially differences in athletic ability. Indeed
Rushton provided a rather bizarre evolutionary explanation for athletic
differences between races: "The reason why Whites and East Asians
have wider hips than Blacks, and so make poorer runners is because
they give birth to larger brained babies. During evolution, increasing
cranial size meant women had to have a wider pelvis. Further, the hor-

mones that give Blacks an edge at sports makes them restless in school and prone to crime."[40] Rushton's racial prejudice drove him to believe that blacks are not only less intelligent, but also genetically predisposed toward criminality.

The evolutionary explanation that Rushton proffered is basically recycled early twentieth-century Nordic racist ideology, though he adorned it with updated language from evolutionary psychology. He argued that as humans migrated out of Africa to Europe and Asia, they faced harsher environments. In order to survive the harsher struggle for existence, they evolved higher intelligence and greater social instincts. Their evolutionary strategy shifted from having as many offspring as possible—which the Africans in the tropics allegedly continued to pursue—to a strategy of having fewer offspring but spending more time and effort caring for them. Notice that thereby Rushton was perpetuating the racial stereotypes of Africans as sexually voracious and less responsible parents than whites or east Asians.[41] In any case, Rushton clearly invoked Darwinian theory to buttress his racist ideology.

MacDonald, a psychology professor at California State University, Long Beach, until his retirement in 2014, also called upon evolutionary psychology to promote his white nationalist views. He focused much of his academic career on analyzing—and justifying—anti-Semitism. He argued that Jews and non-Jews are locked in the Darwinian struggle for existence, and anti-Semitism is an evolutionary strategy for non-Jewish white people to triumph over Jews in that struggle. Many white nationalists express appreciation for MacDonald's work. Devlin, after recommending that fellow white nationalists read all of MacDonald's writings, stated, "I have read Dr. MacDonald's trilogy from beginning to end twice, parts of it more often. I think the interpretation of 'cultures' as group evolutionary strategies will prove a crucial insight."[42] In an article at the neo-Nazi Daily Stormer, Andrew Anglin asserted, "Dr. Kevin MacDonald's work examining the racial nature of Jews is considered crucial to understanding what the Alt-Right is about."[43]

MacDonald published three major books laboriously explaining his belief that the behavior of Jews and anti-Semites has been shaped by the Darwinian struggle for existence. The first of his trilogy was *A People That Shall Dwell Alone: Judaism as a Group Evolutionary Strategy*. In this work he tried to demonstrate that Judaism was a religion that benefitted the Jews by helping them outcompete other ethnic and racial groups in the struggle for existence. The second book, *Separation and Its Discontents: Toward an Evolutionary Theory of Anti-Semitism*, examined the flip side of this, interpreting anti-Semitism as an evolutionary strategy by non-Jews to prevent the Jews from triumphing in the struggle for existence. By taking this stance, MacDonald effectively justified anti-Semitism, as his critics correctly pointed out. MacDonald tried to forestall such criticisms with a rather bizarre and circular argument. He claimed that attacking his work as anti-Semitic was "expectable and completely in keeping with the thesis of this essay."[44] In other words, he implied that anyone complaining about his work being anti-Semitic is either a Jew or a lackey of the Jews.

MacDonald's third and final volume on the Jews is *The Culture of Critique: An Evolutionary Analysis of Jewish Involvement in Twentieth-Century Intellectual and Political Movements*. This work examined the influence of Jewish intellectuals in the twentieth century, especially their contribution to leftist political movements and their rejection of racist ideologies. Among other things, this book regurgitated the hackneyed trope of the Jews as responsible for communism and socialism. In the preface he briefly explained the thesis of his first two books and then intoned, "Ethnic conflict is a recurrent theme throughout the first two volumes, and that theme again takes center stage in this work."[45] The ethnic conflict he is featuring, of course, is between Jews and non-Jews. Indeed, MacDonald interpreted the struggle over ideas as part of the struggle for existence. He stated, "No evolutionist should be surprised at the implicit theory in all this, namely, that intellectual activities of all types may at bottom involve ethnic warfare.... The truly doubtful proposition for an evolutionist is whether real social science as a disinterested attempt to understand human behavior is at all possible."[46] I do not know if MacDonald recognized it (I suspect he did), but this last sentence undermined his own attempts at social science, because apparently MacDonald's own psychological theories have no claim on being "truth," but are merely weapons in "ethnic warfare."

In any case, one point that MacDonald hammered on—and that recurs repeatedly in many other white nationalist writings—is that Jewish intellectuals led the way in undermining racism in Western societies in the early to mid-twentieth century. He disparaged Franz Boas, probably the most influential American anthropologist in the early to mid-twentieth century, depicting him as an archenemy of the (non-Jewish) white race. Boas was a German-born Jew who used his position at Columbia University to bring the idea of cultural determinism to dominance in anthropology and other social sciences. In his chapter on "The Boasian School of Anthropology and the Decline of Darwinism in the Social Sciences," MacDonald slammed Boas for resisting Darwinian and biological explanations for human behavior. He also lamented Boas's influence on reducing racism in American thought.

Toward the end of the book MacDonald included a prediction that seems also to be a plea: "European-derived peoples of the United States will become increasingly unified." This seems to be code for: they will embrace white nationalism. He further predicted that European-derived groups "will develop a united front and a collectivist political orientation vis-à-vis the other ethnic groups. Other groups will be expelled if possible or partitions will be created."[47] MacDonald apparently thought that evolutionary pressures will finally arouse whites to resist other races and take measures to defeat them in the struggle for existence. In an article on a white nationalist website, he emphasized one such measure, immigration restrictions to keep America as white as possible. Indeed, in that article MacDonald argued that if Americans had not abandoned

Darwinian explanations in the social sciences, immigration restrictions would not have been loosened in the mid-twentieth century.[48]

MacDonald not only relied heavily on Darwinism to justify his anti-Semitic white nationalist ideology, but he also bashed intelligent design theory and creationism, casting them as part of a Jewish plot to undermine Darwinism. In a review of the movie *Expelled: No Intelligence Allowed*, MacDonald claimed that by rejecting Darwinism, the Jewish actor Ben Stein, who is featured in the movie, was simply carrying on with the "mainstream of Jewish opinion." In this review MacDonald reiterated his attack on Boas for "more or less obliterating what had been a thriving Darwinian intellectual milieu."[49] Since Ben Stein interviewed me in that film about my book, *From Darwin to Hitler: Evolutionary Ethics, Eugenics, and Racism in Germany*, I suppose that makes me—according to MacDonald's rather bizarre interpretation—a stooge for the Jews.

MacDonald's position ignores the fact that many Jews accept Darwinian evolution and many non-Jews reject it. Also, long before Boas, many non-Jews rejected racism. Further, Boas himself did not reject Darwinism, and many anti-racists are full-fledged adherents to Darwinian theory. MacDonald doubted the ability of social science to be "a disinterested attempt to understand human behavior." Here at least he appears to provide one article of evidence in favor of that conclusion, for the glaring flaws in his reasoning well support the conclusion that his case is strongly motivated not by an objective pursuit of the evidence but instead by his racial animus.

Darwinian Influences on the Alt-Right

The term "alt-right" has come to have a pretty elastic meaning, with the popular news media applying it to all sorts of groups. Here we focus on the core of those who originally called themselves the alt-right. Many of the leading personalities in this movement—and Kevin MacDonald is one of them—promote Darwinism as an integral element of their racist worldview. The alt-right is generally understood to be a movement of white Americans that embraces human inequality, especially racial inequality, while promoting the idea that one should identify with and promote the interests of one's own racial group.[50] In many respects the alt-right is simply a repackaging of neo-Nazism and white nationalism. From my survey of the most influential alt-right online periodicals, it is obvious that Darwinism serves the same ideological function for them that it did for Nazis, neo-Nazis, and earlier white nationalists. It plays a central role in justifying their views on racial inequality and eugenics.

In 2016–17 one of the most publicized figures in the alt-right movement was Richard Spencer, founder of the online Radix Journal. In his speeches Spencer sometimes quotes Nazi propaganda in the original German, and some members of his audience have been known to give Nazi salutes. In a 2017 essay co-authored with F. Roger Devlin, Spencer insisted that racial differences emerged because of evolutionary processes. He outlined the basic evolutionary story of humans originating in Africa and then migrating to Europe and Asia. He argued that "group differences exist as consequences of evolution by natural selection" and "racial differences are a natural and normal consequence of human evolution." He also asserted, "The preference for one's own race is a product of our evolutionary history." Thus, in Spencer's view, white nationalism is hard-wired in our biology. Spencer also embraces the Darwinian explanation so prevalent among Nordic racists and in alt-right circles, stating, "The higher intelligence and lower crime rates of Whites and East Asians as compared with Africans may be due in large part to the selective pressure of cold winters."[51]

Spencer's Radix Journal also publishes numerous articles promoting Darwinian evolution as the basis for their so-called "racial realism." In explaining the fundamentals of the alt-right, Guilluame Durocher wrote, "What is our reason? We believe in Darwin and evolutionary science. Man is, at bottom, a biological entity and, in particular, his potentialities are circumscribed by his genetic heritage. This must be recognized so life may continue its upward evolution, towards the stars, rather than back into the muck." Durocher then criticized liberals for inconsistency, because they say they believe in Darwinism, but then refuse to apply it to public policy.[52]

In a different white nationalist venue, the Occidental Observer, Durocher expressed sympathy with the Third Reich, noting, "The National Socialists proposed a total reformation of society around biocentric norms. This was based on the revolutionary insights of Darwin (which Hitler himself compared to the Copernican revolution), which revealed the natural evolutionary forces which had shaped all life, including all human life." Durocher calls Nazism "a self-conscious group evolutionary strategy (GES) designed to further the interests [of] a genetically-defined German people," and he seems to revel in Nazi atrocities. He states, "The National Socialists observed that in Nature, violence is absolutely fundamental to the survival and development of life, and they sought to be in harmony with this cosmic reality." Durocher argues—as I have in the earlier sections of this book—that Darwinism is a central, defining feature of Nazi ideology.[53]

In 2016 an anonymous article in Spencer's Radix Journal promoted the Darwinian-inspired ideas of the early twentieth-century racist ideologue Madison Grant. The author emphasized the social Darwinist underpinnings of Grant's ideology and recommended his book, Conquest of a Continent, which "is a grand vision of bio-cultural struggle and evolution, in which demography comes alive." The author also promotes Darwinism as the key to promoting white nationalism when he writes, "Darwinism offers a compelling and rational justification for Whites to act on behalf of their ancestors and progeny and feel a shared sense of destiny with their extended kin group.... Darwinism is seemingly more 'effective in rallying Whites, especially elite Whites, than religious feelings.'"[54]

Another prominent figure in the alt-right movement is Jared Taylor, who strongly influenced Spencer. Taylor has also published articles clearly expressing the centrality of Darwinism in his worldview. In 2019 the white nationalist website American Renaissance republished a 1992 essay by Taylor, where he positively reviews Roger Pearson's Race, Intelli-

gence and Bias in Academe. At the top of the article is a picture of Charles Darwin, and the essay begins:

> The discovery of genetics and the development of the theory of evolution were two of the most potentially far-reaching scientific advances of all time. By the turn of the century, thanks to the work of Gregor Mendel (1822–1884) and Charles Darwin (1808–1882), man for the first time had the knowledge with which to direct his own biological destiny. Rather than leave his further development to the genetic accidents that had governed it for millions of years, he could consciously and deliberately improve his very nature.[55]

Taylor not only embraces a social Darwinist version of racism, but he also endorses Darwinian-inspired eugenics. Also, in 2019 American Renaissance republished one of Taylor's 2006 book reviews, where he endorses Richard Lynn's *Race Differences in Intelligence: An Evolutionary Analysis.* According to Taylor, "Prof. Lynn argues that it was the demands of colder, non-African environments that forced the pace of evolution in intelligence and gave rise to race differences."[56] As we have already seen, this is a standard theme of Nordic racist ideology, which underpins their mistaken view that blacks are intellectually inferior to white people. In 2009 Taylor approvingly reviewed the book *Erectus Walks amongst Us* by Richard C. Fuerle. Taylor calls this book "a primer on evolution and genetics, a catalog of how populations differ, an introduction to sociobiology and the concept of genetic interests, and a plea for white survival." Taylor endorsed Fuerle's evolutionary analysis of Africans' inherent biological tendency toward crime and sociopathy. For example, Taylor explains, "In the tropics, where mother and child had a better chance of surviving, it would have been maladaptive not to rape. This may explain high rates of rape among African populations."[57] (This is another example of white nationalists cherry-picking data; it is true that a few African countries have high incidences of rape; however, some African countries, such as Kenya and Uganda, have significantly lower rates than Sweden or Germany or Norway.[58])

Other alt-right figures, such as John Derbyshire, Steve Sailer, and many others, have published articles lauding Darwinism in alt-right websites and periodicals, such as VDARE, Taki's Magazine, the Occidental Observer, and American Renaissance. These carry titles such as "Darwin on the Rise and Fall of the Human Races," "The Evolution of Racial Differences in Morality," and "Ed Dutton with an Evolutionary Perspective on the Rape of Finland."[59] The alt-right proponent Frank Hilliard has even published an essay entitled "The Alternative Right Belongs to the Darwinians" on the Council of European Canadians website.[60] Not all alt-right essays with Darwinian content have such obvious titles, but many articles on these websites recycle the evolutionary ideas I have already explained above.

Conclusion

THE OVERLAPPING categories of neo-Nazis, white nationalists, and alt-right proponents regularly invoke Darwinism to try to demonstrate that their case for racial inegalitarianism and white superiority is scientific. Indeed, by their own admission, Darwinism is central to their ideology, and they regularly ridicule creationists and intelligent design advocates as ignorant. They believe that racial differences have been shaped by natural selection in the struggle for existence. Just like early twentieth-century Nordic racists, some of whom they still promote, they claim that the harsher environment in Europe and Asia provided selective pressure that caused Europeans and Asians to become more intelligent and more cooperative than black Africans. They want to promote further human evolution by practicing racial segregation and introducing eugenics policies. Many of them see nature as their god, and promoting evolution as their god's highest command. Darwinism is thus not only crucial to their explanation of how the world is today, but it is central to their vision of morality and public policy.

9. Conclusion

MOST OBJECTIONS TO THE HISTORICAL CONNECTION BETWEEN Darwinism, racism, and Nazi ideology seem to come from those desperate to protect Darwinism from any such unsavory associations. They ignore masses of data while latching onto small tidbits that seem to corroborate the view they like. Unfortunately for my critics, this book contains a mass of evidence to show that Darwinism did in fact exert a significant influence on Nazi ideology and the many policies based on it.

To be sure, many non-Darwinian elements were synthesized with Darwinism: Aryan supremacy, anti-miscegenation, anti-Semitism, and many more besides. Nonetheless, Nazi racial ideology integrated all these factors into a worldview that stressed the transmutation of species, the evolutionary formation of the human races, the need to advance human evolution, the inevitability of the human struggle for existence, and the need to gain living space (*Lebensraum*) to succeed in the evolutionary struggle. Otto Reche, in the conclusion of his essay on human evolution, connected the dots between evolutionary ideology and the Nazi praxis he championed:

> To sum up: All the mentioned events in the origin of humans and the cultivation of his races can be explained genetically. Without the emergence of hereditary differences, without selection and elimination it could never have come to the formation of highly evolved races and tribes able to accomplish the highest tasks, and never to a higher human culture.
>
> This knowledge obligates us to take up the task, in conscious selection to provide for a racially fit, hereditarily healthy, culture-creating people of the future and thus for the preservation of everything that lifts the life of humans above that of the animals.[1]

Reche's vision was shared by the Nazi regime, whose policies ultimately aimed to improve the human species biologically by advancing human evolution.

Indeed, the historical evidence is overwhelming that human evolution was an integral part of Nazi racial ideology. It held a prominent place in the Nazi school curriculum and in training courses in the Nazi worldview. Nazi officials and SS anthropologists agreed that humans, including the Aryan or Nordic race, had evolved from primates. They believed the Nordic race had evolved to a higher level of intelligence, physical prowess, and social solidarity than other races due to their having faced what evolutionists today would call greater selective pressure. They claimed the stiff selection was caused by the Ice Ages, which had weeded out the weak and sickly, leaving only the brightest and best to propagate the Nordic race. They saw eugenics and racial policy as a means to help the Nordic race evolve to even greater heights.

Moreover, these ideas were not just "Nazi ideas," but were in line with the thinking of many of the leading German biologists and anthropologists before the Nazis came to power. German scientists during the Third Reich recognized the Darwinian thrust of Nazi ideology. The geneticist Otmar von Verschuer and the anthropologist Otto Aichel wrote in a volume celebrating Eugen Fischer's career: "The Führer Adolf Hitler is the first in world history to translate the knowledge about the biological foundations for the evolution of peoples—race, heredity, and selection—into practice."[2]

The British anthropologist Sir Arthur Keith also recognized Hitler's commitment to evolution. In his 1946 book *Ethics and Evolution* he wrote, "The leader of Germany is an evolutionist not only in theory, but, as millions know to their cost, in the rigor of its practice. For him the national 'front' of Europe is also the evolutionary 'front'; he regards himself, and is regarded, as the incarnation of the will of Germany, the purpose of that will being to guide the evolutionary destiny of its people."[3]

A final challenge to my thesis is not to its truth but to its present relevance. According to this objection, most Darwinists today are not racists, so the connection between Darwinism and Nazism is irrelevant.

This objection seems to assume I am making a philosophical argument. I am making a historical one. Intellectual historians are supposed to show the way that ideas influence society and politics (and vice versa). Indeed, many historians who specialize in Nazism and the Third Reich concede that social Darwinism was a major component of Nazi ideology. This is actually a commonplace assertion among experts in Nazism.

If Schopenhauer influenced Hitler (and it seems clear he did), then it is not irrelevant for a historian trying to explain Hitler's worldview to analyze this. Philosophers might still argue about whether Hitler properly understood Schopenhauer, or whether Schopenhauer's philosophy logically led in the direction of Nazism. It is perfectly acceptable to pursue those questions. However, if philosophers deny that Schopenhauer influenced Nazism simply because they want to rescue Schopenhauer from the unsavory reputation of the Nazis, this is unwarranted.

Further, if one is horrified by how the Nazis interpreted Darwinism, and if one wants to show that they were mistaken, then one's argument is not with me, but with the Nazis, whose position I have explained. Take it up with them, they who insisted at numerous points and in venues high and low that their debt to Darwin's theory was great. Also take it up with most late nineteenth and early twentieth-century Darwinian biologists and anthropologists, including Darwin himself, who saw Darwinism as an explanation for racial differences (real or imagined) and as a ground for racist attitudes. In sum, if Darwinism really is not inherently racist, then it is incumbent on Darwinists to explain why it is not.

The matter is also made urgently relevant by the fact that, as shown in Chapter 8, there are white nationalists today who assure us that Darwinism does imply racism, and they are using the same arguments that Darwin, Haeckel, and many other nineteenth and early twentieth-century Darwinists used.

They argue that races are unequal because the races have evolved differently. Of course, conveniently these white nationalists purport to have discovered that their own ancestors—white Europeans—have evolved greater intellectual capacities than other races.

One should not comfort oneself with the notion that such radical ideas are wholly absent from the mainstream academe. That is unlikely. The best one can say with confidence is that these racist ideas are still taboo there. When the Nobel Prize-winning biologist James Watson suggested in 2007 that some racial groups, such as black Africans, had lower intelligence because of their evolutionary history, he faced outrage and sustained criticism.

However, some worrying signs are emerging that the taboo may be cracking. The journal *Evolutionary Psychological Science*, which has eminent evolutionary psychologists, such as Harvard's Steven Pinker, on its editorial board, recently carried an article[4] defending the anti-Semitic, racist views of Kevin MacDonald, the white supremacist and emeritus professor of psychology at California State University, Long Beach, discussed in Chapter 8. As we saw, MacDonald's views are eerily similar to those of scientists that I examine in my historical scholarship. He argues that racial groups are locked in a struggle for existence, behavioral traits are biologically innate, and stereotypically Jewish traits are evolutionary strategies for beating other races in racial competition. MacDonald claims that anti-Semitism is a defensive strategy to help white Europeans and their descendants triumph over the Jews.

Darwin and many early Darwinists saw racism, human inequality, and the struggle for existence as part and parcel of their theory. MacDonald is trying to resurrect this troubling legacy of Darwinian theory, and a mainstream psychology journal chose to give his efforts an assist. If we ignore the legacy of Darwinian racism, civilization may not, as George Santayana warned, be doomed to repeat the history of Nazi Germany. But to borrow from another famous quip about history, it may rhyme with it. Surely there is better poetry we could offer our children.

ENDNOTES

INTRODUCTION

1. I discuss all these themes in depth in *From Darwin to Hitler: Evolutionary Ethics, Eugenics, and Racism in Germany* (New York: Palgrave Macmillan, 2004).

2. Ernst Haeckel, *Die Lebenswunder: Gemeinverständliche Studien über Biologische Philosophie* (Stuttgart: Alfred Kröner, 1904), 450. Throughout the present volume German-to-English translations are mine unless otherwise noted.

3. Many scholars have noted the importance of social Darwinism in Hitler's world view: Richard Weikart, *Hitler's Ethic: The Nazi Pursuit of Evolutionary Progress* (New York: Palgrave Macmillan, 2009); Ian Kershaw, *Hitler*, 2 vols. (New York: Norton, 1998–2000), 2:xli; see also 1:290, 2:19, 208, 405, 780; Richard J. Evans, *The Coming of the Third Reich* (New York: Penguin, 2004), 34–35, and *Third Reich in Power* (New York: Penguin, 2005), 4, 708; Eberhard Jäckel, *Hitler's World View: A Blueprint for Power* (Cambridge, MA: Harvard University Press, 1981), chap. 5; Mike Hawkins, *Social Darwinism in European and American Thought, 1860–1945: Nature as Model and Nature as Threat* (Cambridge: Cambridge University Press, 1997), 277–278; Rainer Zitelmann, *Hitler: Selbstverständnis eines Revolutionärs* (Hamburg: Berg, 1987), 15, 466; and Neil Gregor, *How to Read Hitler* (New York: Norton, 2005), 40. For a longer list, see Weikart, *Hitler's Ethic*, 205–206, n. 6.

4. George L. Mosse, *The Crisis of German Ideology: Intellectual Origins of the Third Reich* (New York: Dunlop and Grossett, 1964), 103.

5. Peter Bowler, "Darwin's Originality," *Science* 323 (January 9, 2009): 226; Michael Ruse, "The Impact of Darwinism," interview by Christopher Fish, *Stanford Review*, April 21, 2008, https://stanfordreview.org/impact-darwinism/.

6. Daniel Gasman, *The Scientific Origins of National Socialism: Social Darwinism in Ernst Haeckel and the German Monist League* (London: MacDonald, 1971), 173.

7. Anne Harrington, *Reenchanted Science: Holism in German Culture from Wilhelm II to Hitler* (Princeton: Princeton University Press, 1996), 262, n.2; Werner Maser takes a similar position in *Hitlers Briefe und Notizen. Sein Weltbild in handschriftlichen Dokumenten* (Düsseldorf: Econ Verlag, 1973), 301.

8. Robert J. Richards, "That Darwin and Haeckel Were Complicit in Nazi Biology," in *Galileo Goes to Trial and Other Myths about Science and Religion*, ed. Ronald Numbers (Cambridge, MA: Harvard University Press, 2009), 177. Richards goes even further, implausibly arguing that Hitler and the Nazis completely rejected biological evolution, in Robert J. Richards, *Was Hitler a Darwinian?: Disputed Questions in the History of Evolutionary Theory* (Chicago: University of Chicago Press, 2013).

9. Richard Weikart, *From Darwin to Hitler: Evolutionary Ethics, Eugenics, and Racism in Germany* (New York: Palgrave Macmillan, 2004), 4–5.

10. On the influence of Mendelism on Nazism, see Amir Teicher, *Social Mendelism: Genetics and the Politics of Race in Germany, 1900–1948* (Cambridge: Cambridge University Press,

2020); see also my review of Teicher's book at *Francia Recensio* 2021, no. 2, https://doi.org/10.11588/frrec.2021.2.81999.

11. Richard Weikart, *From Darwin to Hitler: Evolutionary Ethics, Eugenics, and Racism* (New York: Palgrave Macmillan, 2004), discusses Woltmann, Schemann, and other German thinkers who synthesized Gobineau and Darwinism.

12. Hans-Walter Schmuhl, "Eugenik und Rassenanthropologie," in Robert Jütte et al., *Medizin und Nationalsozialismus: Bilanz und Perspektiven der Forschung* (Göttingen: Wallstein Verlag, 2011), 31. Robert Richards' claim in *Was Hitler a Darwinian?* that the Nazi embrace of Gobineau proves that Nazis were anti-Darwinian is untenable.

13. Richard Weikart, "The Impact of Social Darwinism on Anti-Semitic Ideology in Germany and Austria, 1860–1945," in *Jewish Tradition and the Challenge of Evolution*, eds. Geoffrey Cantor and Marc Swetlitz (Chicago: University of Chicago Press, 2006), 93–115.

14. The best treatment of the neo-Darwinian synthesis in Germany is Thomas Junker, *Die zweite Darwinsche Revolution: Geschichte des synthetischen Darwinismus in Deutschland 1924 bis 1950* (Marburg: Basilisken-Presse, 2004).

15. Christopher Hutton, *Race and the Third Reich: Linguistics, Racial Anthropology and Genetics in the Dialectic of Volk* (Cambridge, UK: Polity, 2005), 212.

1. THE RACISM OF DARWIN AND DARWINISM

1. Charles Darwin to William Graham, July 3, 1881, Darwin Correspondence Project, Letter no. 13230, University of Cambridge, https://www.darwinproject.ac.uk/letter/?docId=letters/DCP-LETT-13230.xml. Letter quoted in Francis Darwin, *Charles Darwin: His Life Told in an Autobiographical Chapter, and in a Selected Series of His Published Letters* (London: Murray, 1902), 64.

2. Charles Darwin, *The Descent of Man*, 2 vols. [1871] (Princeton: Princeton University Press, 1981), 1:201.

3. Most scholars agree that racial struggle is an integral part of Darwin's account of human evolution, and some even explicitly discuss the role of racial extermination in his theory—see Adrian Desmond and James Moore, *Darwin* (New York: Joseph, 1991), xxi, 191, 266–268, 521, 653; Robert M. Young, "Darwinism Is Social," in *The Darwinian Heritage*, ed. David Kohn (Princeton: Princeton University Press, 1985), 609–638; John C. Greene, "Darwin as Social Evolutionist," in *Science, Ideology, and World View: Essays in the History of Evolutionary Ideas* (Berkeley: University of California Press, 1981); Peter Bowler, *Evolution: The History of an Idea*, rev. ed. (Berkeley: University of California Press, 1989), 301; Gregory Claeys, "The 'Survival of the Fittest' and the Origins of Social Darwinism," *Journal of the History of Ideas* 61 (2000): 223–240. A few scholars, however, emphasize Darwin's abolitionist sentiments and sympathy for other races, e.g., Greta Jones, *Social Darwinism and English Thought: The Interaction between Biological and Social Theory* (Sussex: Harvester Press, 1980), 140; Paul Crook, *Darwinism, War and History: The Debate over the Biology of War from the 'Origin of Species' to the First World War* (Cambridge: Cambridge University Press, 1994), 25–28.

4. Desmond and Moore, *Darwin*, xxi.

5. I should note that *Darwin's Sacred Cause* is controversial among historians, and most do not agree with its thesis, because even though Darwin's hatred for slavery is indisputable, it likely had little impact on the formulation of his evolutionary theory.

6. Adrian Desmond and James Moore, *Darwin's Sacred Cause: How a Hatred of Slavery Shaped Darwin's Views on Human Evolution* (Boston: Houghton Mifflin Harcourt, 2009), 149–151.

7. Charles Darwin to Charles Kingsley, February 6, 1862, Darwin Correspondence Project, Letter no. 3439, University of Cambridge, https://www.darwinproject.ac.uk/letter/DCP-LETT-3439.xml. Letter quoted in Desmond and Moore, *Darwin's Sacred Cause*, 318.

8. Christopher Petrusic, "Violence as Masculinity: David Livingstone's Radical Racial Politics in the Cape Colony and the Transvaal, 1845–1852," *International History Review* 26, 1 (2004): 20–55

9. Petina Gappa, "What Nobody Told Me about the Legendary Explorer David Livingstone," *Financial Times Magazine*, February 21, 2020, https://www.ft.com/content/431696a2-52ac-11ea-90ad-25e377c0ee1f.

10. Joshua M. Hall, "Questions of Race in J. S. Mill's Contributions to Logic," *Philosophia Africana* 16, no. 2 (November/December 2014): 73–93.

11. Michael Flannery, *Intelligent Evolution: How Alfred Russel Wallace's World of Life Challenged Darwinism* (Nashville, TN: Erasmus Press, 2020), 54, 65.

12. Charles Darwin, *The Autobiography of Charles Darwin, 1809–1882*, ed. Nora Barlow (New York: Norton, 1958), 76.

13. Charles Darwin, *The Voyage of the Beagle*, ed. Leonard Engel (Garden City, NY: Anchor Books, 1962), 70–71, 208; quotes at 501, 213. For more on Darwin's view of the Australian aborigines, see Barry W. Butcher, "Darwinism, Social Darwinism and the Australian Aborigines: A Reevaluation," in *Darwin's Laboratory: Evolutionary Theory and Natural History in the Pacific*, eds. Roy MacLeod and Philip F. Rehbock (Honolulu: University of Hawaii Press, 1994), 371–394.

14. Darwin, *Voyage of the Beagle*, 433–434.

15. Darwin, *Voyage of the Beagle*, 502.

16. Charles Darwin, *Charles Darwin's Notebooks, 1836–1844: Geology, Transmutation of Species, Metaphysical Inquiries*, eds. Paul Barrett et al. (Ithaca: Cornell University Press, 1987), 414.

17. Darwin, *Charles Darwin's Notebooks*, 537.

18. Darwin, *Descent*, 1:1.

19. In *The Descent of Man* chapter "On the Races of Man," Darwin confirmed his belief that human races differ considerably, not only physically, but also in their mental capacities. For this reason, he considered races to be distinct sub-species. Darwin, *Descent*, 1:216, 227.

20. Charles Darwin, *The Origin of Species* [1859] (London: Penguin, 1968), quotes at 129, 459.

21. Darwin, *Descent*, 1:3.

22. Darwin, *Descent*, 1:35.

23. Darwin, *Descent*, 1:109–110

24. Darwin, *Descent*, 145–146.

25. Daniel Graham, "A Bigger Brain Is Not Better," *Psychology Today*, March 9, 2021, https://www.psychologytoday.com/us/blog/your-internet-brain/202103/bigger-brain-is-not-necessarily-better.

26. Darwin, *Descent*, 1:201.

27. Darwin, *Descent*, 1:238.

28. For a more detailed discussion of Haeckel and other German Darwinists, see my earlier book, *From Darwin to Hitler: Evolutionary Ethics, Eugenics, and Racism in Germany* (New York: Palgrave Macmillan, 2004).

29. Ernst Haeckel, *Natürliche Schöpfungsgeschichte* (Berlin: Georg Reimer, 1868), frontispiece, 555. All translations from the German are mine unless otherwise noted.

30. Ernst Haeckel, *Die Lebenswunder: Gemeinverständliche Studien über Biologische Philosophie* (Stuttgart: Alfred Kröner, 1904), 451–452.

31. Haeckel, *Natürliche Schöpfungsgeschichte*, 218–219, 520.

32. Haeckel, *Natürliche Schöpfungsgeschichte*, 206.

33. Ernst Haeckel, *Ewigkeit. Weltkriegsgedanken über Leben und Tod, Religion und Entwicklungslehre* (Berlin: Georg Reimer, 1917), 35–36, 110–111, 120–123; quotes at 85–86, 35. Emphasis in original. Throughout the present book, emphasis is in the original unless otherwise noted.

34. Friedrich Ratzel, *Sein und Werden der organischen Welt* (Leipzig, 1869), 469.

35. Friedrich Ratzel to Eisig, May 20, 1885, quoted in Gerhard H. Müller, *Friedrich Ratzel (1844–1904): Naturwissenschaftler, Geograph, Gelehrter* (Stuttgart, 1996), 74; see also Woodruff D. Smith, *The Ideological Origins of Nazi Imperialism* (New York: Oxford University Press, 1986), chap. 5; and Mark Bassin, "Imperialism and the Nation State in Friedrich Ratzel's Political Geography," *Progress in Human Geography* 11 (1987): 473–495.

36. Friedrich Ratzel, *Politische Geographie oder die Geographie der Staaten, des Verkehres und des Krieges*, 2nd ed. (Munich, 1903), 44, 129–153, 371–374; quotes at 153, 143.

37. On the prevalence of scientific racism, see Richard Weikart, *From Darwin to Hitler: Evolutionary Ethics, Eugenics, and Racism in Germany* (New York: Palgrave Macmillan, 2004); Nancy Stepan, *The Idea of Race in Science: Great Britain, 1800–1960* (New York: Archon Books, 1982); and Benoit Massin, "From Virchow to Fischer: Physical Anthropology and 'Modern Race Theories' in Wilhelmine Germany," in George W. Stocking, ed., *Volksgeist as Method and Ethic* (Madison: University of Wisconsin Press, 1996), 79–154.

2. Hitler's Darwinian Worldview

1. Traudl Junge, *Bis zur letzten Stunde: Hitlers Sekretärin erzählt ihr Leben*, ed. Melissa Müller (Munich: Claassen Verlag, 2002), 122.

2. Wilfried Daim, *Der Mann, der Hitler die Ideen Gab: Jörg Lanz von Liebenfels*, 3rd ed. (Vienna: Ueberreuter, 1994).

3. For further discussion of influences on Hitler, see Richard Weikart, *Hitler's Religion: The Twisted Beliefs That Drove the Third Reich* (Washington, DC: Regnery History, 2016).

4. For information on the books in Hitler's library, see Phillip Gassert and Daniel S. Mattern, eds., *The Hitler Library: A Bibliography* (Westport, CT: Greenwood Press, 2001).

5. Fritz Lenz, *Die Rasse als Wertprinzip, Zur Erneuerung der Ethik* (Munich: J. F. Lehmann, 1933), quotes at 39, 6–7. Hitler's library had the original 1917 version.

6. Erwin Baur, "Der Untergang der Kulturvölker im Lichte der Biologie," *Deutschlands Erneuerung* 6 (May 1922): 257–268; quote at 261.

7. Hans F. K. Günther, *Rassenkunde des deutschen Volkes*, 3rd ed. (Munich: J. F. Lehmanns Verlag, 1923), 246, 409, 416–419, 502–503; quote at 21. For Günther's views on Darwinism, see also Hans F. K. Günther, *Rassenkunde Europas*, 3rd ed. (Munich: J. F. Lehmanns

Verlag, 1929), 323, and Hans F. K. Günther, *Mein Eindruck von Adolf Hitler* (Pähl: F. von Bebenburg, 1969), 153.

8. Erwin Baur, Eugen Fischer, and Fritz Lenz, *Grundriss der menschlichen Erblichkeitslehre und Rassenhygiene, vol. 1: Menschliche Erblichkeitslehre,* 2nd ed. (Munich: J. F. Lehmanns Verlag, 1923), 131–132, 143, 148.

9. Baur, Fischer, and Lenz, *Grundriss der menschlichen Erblichkeitslehre,* 1: 406–420.

10. Baur, Fischer, and Lenz, *Grundriss der menschlichen Erblichkeitslehre und Rassenhygiene, vol. 2: Menschliche Auslese und Rassenhygiene (Eugenik),* 2nd ed. (Munich: J. F. Lehmanns Verlag, 1923), 166–192.

11. Baur, Fischer, and Lenz, *Grundriss der menschlichen Erblichkeitslehre,* 2: 3.

12. Heiner Fangerau, *Etablierung eines rassenhygienischen Standardwerkes 1921–1941: Der Baur-Fischer-Lenz im Spiegel der zeitgenössischen Rezensionsliteratur* (Frankfurt: Peter Lang, 2001), 87, 251–252.

13. Fangerau, *Etablierung eines rassenhygienischen Standardwerkes,* 250.

14. Fritz Lenz, "Die Stellung des Nationalsozialismus zur Rassenhygiene," *Archiv für Rassen- und Gesellschaftsbiologie* 25, no. 3 (1931): 301–302.

15. Adolf Hitler, "Zukunft oder Untergang," March 6, 1927, microfilm, p. 2, Hoover Institution, NSDAP Hauptarchiv, Reel 2, Folder 59.

16. Adolf Hitler, *Mein Kampf* [1925 and 1927], trans. Ralph Manheim (Boston: Houghton Mifflin, 1943), 383.

17. Ernst Haeckel, *Freie Wissenschaft und freie Lehre* (Stuttgart, 1878), 72–75; Haeckel, "Die Wissenschaft und der Umsturz," *Die Zukunft* 10 (1895): 205–206.

18. Hitler, *Mein Kampf,* 144–145.

19. Hitler, *Mein Kampf,* 151–153.

20. Adolf Hitler, *Mein Kampf* [1925 and 1927], trans. Barrows Mussey (New York: Stackpole Sons, 1939), 278.

21. Adolf Hitler, *Hitler's Second Book: The Unpublished Sequel to* Mein Kampf, ed. Gerhard Weinberg (New York: Enigma Books, 2003), 6–7.

22. Adolf Hitler, "War der Zweite Weltkrieg für Deutschland vermeidbar?," May 30, 1942, in *Hitlers Tischgespräche im Führerhauptquartier,* ed. Henry Picker (Frankfurt: Ullstein, 1989), 491.

23. Hitler, *Hitlers Tischgespräche,* 492.

24. Adolf Hitler, "Was ist Nationalsozialismus?," August 6, 1927, in *Hitler: Reden, Schriften, Anordnungen,* 2: 442.

25. Adolf Hitler, *The Speeches of Adolf Hitler, April 1922–August 1939,* ed. Norman H. Baynes, 2 vols. (Oxford: Oxford University Press, 1942), 1:464. Here Hitler did not explicitly say that humans had evolved from apes, but he clearly implied it.

26. Adolf Hitler, "Hitler vor Bauarbeitern in Berchtesgaden über nationalsozialistische Wirtschaftspolitik am 20. Mai 1937," in *"Es spricht der Führer": 7 exemplarische Hitler-Reden,* ed. Hildegard von Kotze and Helmut Krausnick (Gütersloh: Sigbert Mohn Verlag, 1966), 220–221.

27. Hitler, *Hitlers Tischgespräche,* 75.

28. Adolf Hitler, March 1, 1942, in Adolf Hitler, *Monologe im Führerhauptquartier 1941–1944: Die Aufzeichnungen Heinrich Heims*, ed. Werner Jochmann (Hamburg: Albrecht Knaus, 1980), 310.

29. Adolf Hitler, "Ansprache des Führers vor Generalen und Offiziers am 22.6.1944 im Platterhof," microfilm, p. 2, Hoover Institution, NSDAP Hauptarchiv, Reel 2, Folder 51.

30. Hitler, "Ansprache des Führers vor Generalen und Offiziers am 22.6.1944 im Platterhof," 3–4.

31. Hitler, *Hitlers Tischgespräche*, 93–94.

3. EVOLUTIONARY THEORY IN NAZI SCHOOLS

1. Frieda Wunderlich, "Education in Nazi Germany," *Social Research* 4, no. 3 (September 1937): 349.

2. I. L. Kandel, "Education in Nazi Germany," *The Annals of the American Academy of Political and Social Science* 182, no. 1 (November 1935): 159.

3. *Erziehung und Unterricht in der Höheren Schule: Amtliche Ausgabe des Reichs- und Preussische Ministeriums für Wissenschaft, Erziehung und Volksbildung* (Berlin: Weidmannsche Verlagsbuchhandlung, 1938), quotes at 148–149, 141, 157, 160.

4. H. Linder and R. Lotze, "Lehrplanentwurf für den biologischen Unterricht an den höheren Knabenschulen. Bearbeitet im Auftrag des NSLB Reichsfachgebiet Biologie," in *Der Biologe*. This appeared as separate supplement without page numbering in vol. 6 (1937). It appeared immediately after Heft 1 in the copy I saw. An earlier draft of the same document is contained in *Der Biologe* (1936): 239–246.

5. Paul Brohmer, *Der Unterricht in der Lebenskunde*, 4th ed. (Osterwieck-Harz: A. W. Zickfeldt, 1943), ix, 2–3; see also Paul Brohmer, *Biologieunterricht und völkische Erziehung* (Frankfurt: Verlag Moritz Diesterweg, 1933), vi–vii, 3.

6. Ferdinand Rossner, *Der Weg zum ewigen Leben der Natur: Gegenwartsfragen der biologischen Lebenskunde*, 2nd ed. (Langensalza: Verlag von Julius Beltz, 1937), 100.

7. Ferdinand Rossner, ed., *Handbuch für den Biologieunterricht*, 3 vols. (Langensalza: Julius Beltz, 1939–41), 1:2–11, 92, 2:100–101, 114–118, 3:208–209.

8. Sheila Faith Weiss, *The Nazi Symbiosis: Human Genetics and Politics in the Third Reich* (Chicago: University of Chicago Press, 2010), 225.

9. For more on the connections between Marxism and Lamarckism, see Richard Weikart, *Socialist Darwinism: Evolution in Germany Socialist Thought from Marx to Bernstein* (San Francisco: International Scholars Publications, 1998).

10. Jakob Graf, *Biologie für Oberschule und Gymnasium*, vol. 4: *Ausgabe für Knabenschulen* (Munich: J. F. Lehmanns Verlag, 1942), chs. 9–10.

11. Erich Meyer and Karl Zimmermann, *Lebenskunde: Lehrbuch der Biologie für höhere Schulen*, vol. 2 (Erfurt: Verlag Kurt Stenger, 1940), 2:333.

12. Hermann Wiehle and Marie Harm, *Lebenskunde für Mittelschulen*, vol. 6: *Klasse 6 für Jungen* (Halle a.d. Saale: Hermann Schroedel Verlag, 1942), 132.

13. Sepp [Josef] Burgstaller, *Erblehre, Rassenkunde und Bevölkerungspolitik: 400 Zeichenskizzen für den Schulgebrauch* (Vienna: Deutscher Verlag für Jugend und Volk, 1941), 29–31.

14. "Verzeichnis der zur Beschaffung für Schulbüchereien (Lehrer- und Schülerbüchereien) geeigneten Bücher und Schriften," *Deutsche Wissenschaft Erziehung und Volksbildung: Amtsblatt des Reichsministeriums für Wissenschaft, Erziehung und Volksbildung und der*

Unterrichtsverwaltungen der Länder 2 (1936): 276; Richard Hesse, *Abstammungslehre und Darwinismus*, 7th ed. (Leipzig: B. G. Teubner, 1936), 48–55.

15. "Verzeichnis der zur Beschaffung für Schulbüchereien (Lehrer- und Schülerbüchereien) geeigneten Bücher und Schriften," 515; Martin Staemmler, *Rassenpflege und Schule*, 3rd ed. (Langensalza: Hermann Beyer und Söhne, 1937), 13, 32–36.

16. *Der Biologe* 8 (1939).

17. Walter Greite, "Aufbau und Aufgaben des Reichsbundes für Biologie," *Der Biologe* 8 (1939): 233–241.

18. F. Rossner, "Systematik und Entwicklungsgedanke im Unterricht," *Der Biologe* 8 (1939): 366–372; other articles criticizing anti-evolutionism include Ernst Lehmann, "Entwicklungslehre und 'Imprimatur,'" *Der Biologe* 6 (1937): 291–293; O. Schwarz, "Irrtum und Wahrheit in der Biologie. Kritik der Abstammungslehre," *Der Biologe* 6 (1937): 55–58; Ernst Lehmann, "'Entfaltung—nicht Abstammung,'" *Der Biologe* 7 (1938): 45–47; Christian von Krogh, "Immer wieder: Abstammung oder Schöpfung? Eine Weltanschauungsfrage," *Der Biologe* 9 (1940): 414–417; F. Schwanitz, "Ein Kreuzzug gegen die Abstammungslehre," *Der Biologe* 9 (1940): 407–413.

19. Konrad Lorenz, "Nochmals: Systematik und Entwicklungsgedanke im Unterricht," *Der Biologe* 9 (1940): 24–36; quotes at 24, 32.

20. "Richtlinien für die Bestandsprüfung in den Volksbüchereien Sachsens," *Die Bücherei: Zeitschrift für deutsche Schrifttumspflege* 2 (1935): 279–280.

21. Richard Weikart, "'Evolutionäre Aufklärung'? Zur Geschichte des Monistenbundes" in *Wissenschaft, Politik, und Öffentlichkeit: Von der Wiener Moderne bis zur Gegenwart*, eds. Mitchell G. Ash and Christian H. Stifter (Vienna: WUV Universitätsverlag, 2002), 131–148.

22. Guenther Lewy, *Harmful and Undesirable: Book Censorship in Nazi Germany* (New York: Oxford University Press, 2016).

23. "Zu unserm Sonderverzeichnis: Rassenpflege, warum und wie?" *Die Bücherei: Zeitschrift für deutsche Schrifttumspflege* 1, 1 (1934): 46.

24. "Wesentliche Bücher des Jahres 1935," *Die Bücherei: Zeitschrift für deutsche Schrifttumspflege* 3 (1936): 64–65; Alfred Kühn, Martin Staemmler, and Friedrich Burgdörfer, *Erbkunde, Rassenpflege, Bevölkerungspolitik: Schicksalsfragen des deutschen Volkes*, 2nd ed. (Leipzig: Quelle und Meyer, 1935), chap. 7.

25. Günther Tschich, "Die hundert Bücher des Jahres: Zur Verkaufsstatistik des Einkaufshauses für Büchereien," *Die Bücherei: Zeitschrift für deutsche Schrifttumspflege* 4 (1937): 14–20.

26. Review of Edgar Dacqué, *Urweltkunde Süddeutschlands: Mit einer allgemeinen geologischen Einführung* (Munich, 1934), in *Die Bücherei: Zeitschrift für deutsche Schrifttumspflege* 3 (1936): 278; Hermann Eisner, review of Erich Schneider, *Entwicklungsgeschichte der naturwissenschaftlichen Weltanschauung von der griechischen Naturphilosophie bis zur modernen Vererbungslehre*, in *Die Bücherei: Zeitschrift für deutsche Schrifttumspflege* 4 (1937): 67–68; Hermann Propach, review of Theodor Schmucker, *Geschichte der Biologie: Forschung und Lehre* (1936), in *Die Bücherei: Zeitschrift für deutsche Schrifttumspflege* 5 (1938): 626–627; Margarethe Kölle, review of Bruno Gebhard, ed., *Wunder des Lebens*, in *Die Bücherei: Zeitschrift für deutsche Schrifttumspflege* 5 (1938): 620.

4. DARWINIAN SCIENTISTS OF THE THIRD REICH

1. Hans F. K. Günther, *Mein Eindruck von Adolf Hitler* (Pähl: F. von Bebenburg, 1969), 18–21; Hitler to unknown person, February 2, 1930, in *Hitler: Reden, Schriften, Anordnungen, Februar 1925 bis Januar 1933*, vol. 3, part 3: *Januar 1930-September 1930* (Munich: K. G. Saur, 1992-2003), 61–62; see also Uwe Hossfeld, "Die Jenaer Jahre des Rasse-Günther von 1930 bis 1935," *Medizinhistorisches Journal* 34 (1999): 47–103.

2. Hossfeld, "Die Jenaer Jahre des Rasse-Günther," 56.

3. Günther, *Mein Eindruck von Adolf Hitler*, 59, 100–101.

4. Hans F. K. Günther, *Rassenkunde des deutschen Volkes*, 3rd ed. (Munich: J. F. Lehmanns Verlag, 1923), 21–24.

5. Günther, *Rassenkunde des deutschen Volkes*, 246, 279–80.

6. Günther, *Rassenkunde des deutschen Volkes*, chap. 22.

7. Hans F. K. Günther, *Volk und Staat in ihrer Stellung zu Vererbung und Auslese*, 2nd ed. (Munich: J. F. Lehmanns Verlag, 1933), 17–18, 24–26.

8. Michael Hesch, "Otto Reche als Rassenforscher," in *Kultur und Rasse: Otto Reche zum 60. Geburtstag*, eds. Michael Hesch and Günther Spannaus (Munich: J. F. Lehmanns Verlag, 1939), 14.

9. Otto Reche, "Ludwig Woltmann," in *Woltmanns Werke*, vol. 1: *Politische Anthropologie* (Leipzig: Justus Dörner, 1936), 7–8.

10. Otto Reche, "Vorwort des Herausgebers," in *Woltmanns Werke*, vol. 1: *Politische Anthropologie* (Leipzig: Justus Dörner, 1936), 32.

11. For more on Woltmann, see Richard Weikart, *From Darwin to Hitler: Evolutionary Ethics, Eugenics, and Racism in Germany* (New York: Palgrave Macmillan, 2004), 119–122, 196–199.

12. Michael Hesch, "Otto Reche als Rassenforscher," in *Kultur und Rasse: Otto Reche zum 60. Geburtstag*, eds. Michael Hesch and Günther Spannaus (Munich: J. F. Lehmanns Verlag, 1939), 13–14.

13. Werner Kulz, "Die politisch-weltanschauliche Bedeutung der Arbeiten Otto Reches," in *Kultur und Rasse: Otto Reche zum 60. Geburtstag*, 18.

14. Julius Andree, "Mittel- und Westeuropa als älteste Kulturherde der Nordischen Rasse," in *Kultur und Rasse: Otto Reche zum 60. Geburtstag*, 39–50.

15. Quoted in Werner Kulz, "Die politisch-weltanschauliche Bedeutung der Arbeiten Otto Reches," in *Kultur und Rasse: Otto Reche zum 60. Geburtstag*, 21.

16. Otto Reche, "Entstehung des Menschen und seiner Rassen," in *Rassenhygiene für Jedermann*, ed. Ernst Wegner (Dresden: Theodor Steinkopff, 1934), 11–29.

17. For an extensive treatment of Reche's connections to the Nazi regime, see Katja Geisenhainer, *"Rasse ist Schicksal": Otto Reche (1879–1966) – ein Leben als Anthropologe und Völkerkundler* (Leipzig: Evangelische Verlagsanstalt, 2002).

18. Eugen Fischer, *Sozialanthropologie und ihre Bedeutung für den Staat* (Freiburg: Speyer and Kaerner, 1910), 18–19.

19. Eugen Fischer, *Die Rehobother Bastards und das Bastardierungsproblem beim Menschen* (Jena: Gustav Fischer, 1913).

20. Eugen Fischer, "Rassen und Rassenbildung," in *Handwörterbuch der Naturwissenschaften* (Jena: Gustav Fischer, 1912–13), 8: 78–106.

21. Eugen Fischer, *Rasse und Rassenentstehung beim Menschen* (Berlin: Verlag Ullstein, 1927), 34–46.

22. Eugen Fischer, Review of Hans Weinert, *Der geistige Aufstieg der Menschheit* (1940), in *Zeitschrift für Morphologie und Anthropologie* 40 (1943): 188–189.

23. For more on this relationship between Fischer and the regime, see Sheila Faith Weiss, *The Nazi Symbiosis: Human Genetics and Politics in the Third Reich* (Chicago: University of Chicago Press, 2010), especially chap. 2.

24. Reiner Pommerin, *"Sterilisierung der Rheinlandbastarde": Das Schicksal einer farbigen deutschen Minderheit 1918–1937* (Düsseldorf: Droste Verlag, 1979), 49–50, 71–78.

25. Niels Lösch, *Rasse als Konstrukt: Leben und Werk Eugen Fischers* (Frankfurt a.M.: Lang, 1997), 363.

26. "Verzeichnis der zur Beschaffung für Schulbüchereien (Lehrer- und Schülerbüchereien) geeigneten Bücher und Schriften," *Deutsche Wissenschaft Erziehung und Volksbildung: Amtsblatt des Reichsministeriums für Wissenschaft, Erziehung und Volksbildung und der Unterrichtsverwaltungen der Länder* 2 (1936): 64

27. Hans Weinert, *Die Rassen der Menschheit* (Leipzig: B. G. Teubner, 1935), 4.

28. Weinert, *Die Rassen der Menschheit*, 21.

29. Weinert, *Die Rassen der Menschheit*, 136.

30. Hans Weinert, *Biologische Grundlagen für Rassenkunde und Rassenhygiene* (Stuttgart: Ferdinand Enke, 1934), 4.

31. Weinert, *Biologische Grundlagen für Rassenkunde und Rassenhygiene*, 30.

32. Hans Weinert, *Entstehung der Menschenrassen*, 2nd ed. (Stuttgart: Fredinand Enke Verlag, 1941), 314–315.

33. "Wertvolle Bücher," *Neues Volk: Blätter des Rassenpolitischen Amtes der N.S.D.A.P.* 3, 2 (1935): 46–48; "Bücher, die zu empfehlen sind," *Neues Volk* 3, 6 (1935): 46.

34. Heinz Brücher, "Lebenskunde," *Nationalsozialistische Monatshefte* 8 (1937): 190–192.

35. Ute Deichmann in *Biologists under Hitler*, trans. Thomas Dunlap (Cambridge, MA: Harvard University Press, 1996), 270.

36. For a good discussion of Astel, see Uwe Hossfeld, *Geschichte der biologischen Anthropologie in Deutschland: Von den Anfängen bis in die Nachkriegszeit* (Stutgart: Franz Steiner Verlag, 2005), 217, 220, 231–233; Paul Weindling, "'Mustergau' Thüringen: Rassenhygiene zwischen Ideologie und Machtpolitik," in *'Kämperische Wissenschaft': Studien zur Universität Jena im Nationalsozialismus*, eds. Uwe Hossfeld et al (Cologne: Böhlau Verlag, 2003), 1013–1026; and Uwe Hossfeld, "Menschliche Erblehre, Rassenpolitik und Rassenkunde (-biologie) an den Universitäten Jena und Tübingen von 1934–45: Ein Vergleich," in *Ethik der Biowissenschaften: Geschichte und Theorie*, eds. Eve-Marie Engels et al. (Berlin: Verlag für Wissenschaft und Bildung, 1998), 361–392.

37. Karl Astel, "Rassendämmerung und ihre Meisterung durch Geist und Tat als Schicksalsfrage der weissen Völker," *Nationalsozialistische Monatshefte* 6 (1935): 194–195, 202–203; quote at 203.

38. Uwe Hossfeld, *Gerhard Heberer (1901–1973): Sein Beitrag zur Biologie im 20. Jahrhundert* (Berlin: Verlag für Wissenschaft und Bildung, 1997), 63–64.

39. Uwe Hossfeld, *Gerhard Heberer (1901–1973): Sein Beitrag zur Biologie im 20. Jahrhundert* (Berlin: Verlag für Wissenschaft und Bildung, 1997), 60–62, 67, 186–187; Hossfeld,

Geschichte der biologischen Anthropologie in Deutschland, 261; Deichmann, *Biologists under Hitler,* 272–273, 274.

40. Hossfeld, *Gerhard Heberer,* 85–86.

41. Gerhard Heberer, *Rassengeschichtliche Forschungen im indogermanischen Urheimatgebiet* (Jena: Gustav Fischer, 1943), 1–4.

42. Thomas Junker, *Die zweite Darwinsche Revolution: Geschichte des Synthetischen Darwinismus in Deutschland 1924 bis 1950* (Marburg: Basilisken-Presse, 2004), 297.

43. Wilhelm Gieseler, "Die Fossilgeschichte des Menschen," in *Die Evolution der Organismen: Ergebnisse und Probleme der Abstammungslehre,* ed. Gerhard Heberer (Jena: Gustav Fischer, 1943), 615.

44. Junker, *Die zweite Darwinsche Revolution,* 258–259; "Mitteilungen," *Anthropologischer Anzeiger* 10 (1933): 245.

45. Uwe Hossfeld, "Menschliche Erblehre, Rassenpolitik und Rassenkunde (-biologie) an den Universitäten Jena und Tübingen von 1934-45: Ein Vergleich," in Eve-Marie Engels, Thomas Junker and Michael Weingarten, eds., *Ethik der Biowissenschaften: Geschichte und Theorie* (Berlin: Verlag für Wissenschaft und Bildung, 1998), 372–373.

46. Geisenhainer, *"Rasse ist Schicksal": Otto Reche,* 475; "Wissenschaft und Weltanschauung," *Neues Volk* 6, 12 (1938): 27.

47. Walter Gross, "Rassische Geschichtsbetrachtung," in *Europas Geschichte als Rassenschicksal: Vom Wesen und Wirken der Rassen im europäischen Schicksalsraum,* ed. Rolf L. Fahrenkrog (Leipzig: Hesse und Becker Verlag, n.d.), 14.

48. Otto Reche, "Die Genetik der Rassenbildung beim Menschen," in *Die Evolution der Organismen: Ergebnisse und Probleme der Abstammungslehre,* ed. Gerhard Heberer (Jena: Gustav Fischer, 1943), 700.

5. NAZI EUGENICS AND EUTHANASIA

1. I discuss all these points further in Richard Weikart, *Hitler's Ethic: The Nazi Pursuit of Evolutionary Progress* (New York: Palgrave Macmillan, 2009).

2. *Alles Leben ist Kampf* [1937] can be seen at YouTube, video, 9:02, June 14, 2007, https://www.youtube.com/watch?v=jeaIwnNj-QA.

3. *Alles Leben ist Kampf.*

4. Francis Galton, *Memories of My Life* (London: Methuen, 1908), 287.

5. Both quotations are in Michael Bulmer, *Francis Galton: Pioneer of Heredity and Biometry* (Baltimore: Johns Hopkins University Press, 2003), 42.

6. Nicholas Wright Gillham, *A Life of Sir Francis Galton: From African Exploration to the Birth of Eugenics* (Oxford: Oxford University Press, 2001), 328, 197.

7. Charles Darwin, *The Descent of Man* (London: J. Murray, 1871), Part 1, Chapter 5, 168.

8. Darwin, *The Descent of Man,* 168–169.

9. Edward M. East, *Mankind at the Crossroads* (New York: Charles Scribner's Sons, 1924), vi.

10. East, *Mankind at the Crossroads,* 23.

11. Margaret Sanger, *Woman and the New Race* (New York: Eugenics Publishing Company, 1920), 229.

12. Margaret Sanger, *The Pivot of Civilization* (New York: Brentano's, 1922), 170–171.

13. "Certificate of Appreciation from the Second International Congress of Eugenics," Onview: Digital Collections & Exhibits, Countway Library, Harvard University, September 1921, http://collections.countway.harvard.edu/onview/files/original/6a44ea28b6b0d03c0 91cadd39f734c99.jpg.

14. I discuss the Darwinian influences on eugenics in the late nineteenth and early twentieth centuries in greater depth in *From Darwin to Hitler: Evolutionary Ethics, Eugenics, and Racism in Germany* (New York: Palgrave Macmillan, 2004). Therein I include considerable material on Haeckel, Schallmayer, and Ploetz, whom I discuss briefly here.

15. Ernst Haeckel, *Natürliche Schöpfungsgeschichte.* 2nd ed. (Berlin: Georg Reimer, 1870), 152–155.

16. Ernst Haeckel, *Die Lebenswunder: Gemeinverständliche Studien über Biologische Philosophie* (Stuttgart: Alfred Kröner, 1904), 22, 132–136; quote at 445.

17. Wilhelm Schallmayer, *Die drohende physische Entartung der Culturvölker,* 2nd ed. (Berlin: Heuser's Verlag, 1891), 1, 4.

18. Heinrich Ernst Ziegler, "Einleitung zu dem Sammelwerke Natur und Staat," in *Natur und Staat,* vol. 1 (bound with Heinrich Matzat, *Philosophie der Anpassung*) (Jena, 1903), 1–2; Klaus-Dieter Thomann and Werner Friedrich Kümmel, "Naturwissenschaft, Kapital und Weltanschauung: Das Kruppsche Preisausschreiben und der Sozialdarwinismus," *Medizinhistorisches Journal* 30 (1995): 99–143, 205–243. Sheila Faith Weiss provides a good discussion of the Krupp Prize competition in *Race Hygiene and National Efficiency: The Eugenics of Wilhelm Schallmayer* (Berkeley: University of California Press, 1987), 64–74.

19. Wilhelm Schallmayer, *Vererbung und Auslese im Lebenslauf der Völker. Eine Staatswissenschaftliche Studie auf Grund der neueren Biologie* (Jena: Gustav Fischer, 1903), 243, quote at 250.

20. Wilhelm Schallmayer to Alfred Grotjahn, June 3, 1910, in Alfred Grotjahn papers, Humboldt University Archives, Berlin; Schallmayer makes a similar claim in "Rassedienst," *Sexual-Probleme* 7 (1911): 547.

21. Alfred Ploetz to Carl Hauptmann, January 14, 1892, in Carl Hauptmann papers, K 121, Akademie der Künste Archives, Berlin; Alfred Ploetz to Ernst Haeckel, April 18, 1902, in Ernst Haeckel papers, Ernst-Haeckel-Haus, Jena; for more on Ploetz, see W. Doeleke, "Alfred Ploetz (1860–1940): Sozialdarwinist und Gesellschaftsbiologe," dissertation, University of Frankfurt, 1975.

22. Alfred Ploetz to Ernst Haeckel, October 19, 1903, in Alfred Ploetz papers, privately held by Wilfried Ploetz, Herrsching am Ammersee; "Aufforderung zum Abonnement. Archiv für Rassen- und Gesellschafts-Biologie," August 1903, in Felix von Luschan papers (among Alfred Ploetz's letters), in Staatsbibliothek Preussischer Kulturbesitz, Berlin.

23. "Der Sachverständigen-Beirat für Bevölkerungs- und Rassenpolitik," *Archiv für Rassen- und Gesellschaftsbiologie* 27, no. 4 (1934): 419–420.

24. "Ehrung Alfred Ploetz' durch den Führer," *Archiv für Rassen- und Gesellschaftsbiologie* 29, no. 4 (1936): no page number, but it is the first page in the issue.

25. Ian Dowbiggin, *A Merciful End: The Euthanasia Movement in Modern America* (Oxford: Oxford University Press, 2003), 8.

26. N. D. A. Kemp, *"Merciful Release": The History of the British Euthanasia Movement* (Manchester: Manchester University Press, 2002), 19.

27. Hans-Walter Schmuhl, *Rassenhygiene, Nationalsozialismus, Euthanasie: Von der Verhütung zur Vernichtung 'lebensunwerten Lebens' 1890–1945* (Göttingen, 1987), 18–19, quote at 106.

28. Adolf Hitler, *Mein Kampf* [1925 and 1927], trans. Ralph Manheim (Boston: Houghton Mifflin, 1943), 29–30.

29. Jay Joseph and Norbert Wetzell, "Ernst Rüdin: Hitler's Racial Hygiene Mastermind," *Journal of the History of Biology* 46 (2013): 3.

30. Arthur Gütt, Ernst Rüdin, and Falk Ruttke, *Gesetz zur Verhütung erbkranken Nachwuchses vom 14. Juli 1933* (Munich: J. F. Lehmanns Verlag, 1934), 13, 50.

31. Ernst Rüdin, ed., *Erblehre und Rassenhygiene im völkischen Staat* (Munich: J. F. Lehmanns Verlag, 1934), 104–105.

32. L. Plate, "Ein moderner Gegner der Abstammungslehre," *Archiv für Rassen- und Gesellschafts-Biologie* 29 (1935): 249–256.

33. Information on Staemmler's involvement in the Nazi Party is in Hans-Christian Harten, Uwe Neirich, and Matthias Schwerendt, *Rassenhygiene als Erziehungsideologie des Dritten Reichs: Bio-bibliographisches Handbuch* (Berlin: Akademie Verlag, 2006), 284.

34. Ernst Wegner, *Rassenhygiene für Jedermann* (Dresden: Theodor Steinkopff, 1934), published three lectures by Staemmler.

35. "Verzeichnis der Lehrmittel über Erbkunde, Erbpflege, Rassenkunde und Bevölkerungspolitik," *Deutsche Wissenschaft Erziehung und Volksbildung: Amtsblatt des Reichsministeriums für Wissenschaft, Erziehung und Volksbildung und der Unterrichtsverwaltungen der Länder* 3 (1937): 247.

36. Martin Staemmler, *Deutsche Rassenpflege* (n.p.: Tornisterschrift des Oberkommandos der Wehrmacht Abt. Inland, 1941).

37. Martin Staemmler, *Rassenpflege im völkischen Staat* (Munich: J. F. Lehmanns Verlag, 1937), 5.

38. Staemmler, *Rassenpflege im völkischen Staat*, 17–22.

39. Staemmler, *Rassenpflege im völkischen Staat*, 35.

40. Martin Staemmler, *Die Auslese im Erbstrom des Volkes* (Berlin: Zentralverlag der NSDAP., Franz Eher Nachf., 1939), 16–17.

41. Staemmler, *Die Auslese im Erbstrom des Volkes*, 20.

42. Ernst Wegner, ed., *Rassenhygiene für Jedermann* (Dresden: Theodor Steinkopff, 1934), v, 13–17, 161–163.

43. Alfred Kühn, Martin Staemmler, and Friedrich Burgdörfer, *Erbkunde, Rassenpflege, Bevölkerungspolitik. Schicksalsfragen des deutschen Volkes*, 2nd ed. (Leipzig: Quelle und Meyer, 1935), 80–94; quote at 93.

44. Michael Burleigh, *Death and Deliverance: Euthanasia in Germany, 1900–1945* (Cambridge: Cambridge University Press, 1994), 192.

45. Ulf Schmidt, *Medical Films, Ethics and Euthanasia in Nazi Germany: The History of the Medical Research and Teaching Films of the Reich Office for Educational Films/Reich Institute for Films in Science and Education, 1933–1945* (Husum: Matthiesen Verlag, 2002), 137.

46. *Erbkrank*, 1936. This can be viewed on YouTube, video, 11:29, January 14, 2017, https://www.youtube.com/watch?v=QFIqyM92GAI&list=PLfAXBbqlQQS_CIU4CoTwYIwii66ucm1XS&index=3. (The YouTube description incorrectly dates the film as 1934.)

47. *Selling Murder: The Killing Films of the Third Reich*, directed by Joanna Mack, written and researched by Michael Burleigh (London: Domino Films, 1991). It can be viewed at YouTube, video, 53:37, November 30, 2014, https://www.youtube.com/watch?v=c2kV83nPWnM. The *Das Erbe* [1935] section begins at 8:40.

48. *Selling Murder*, at about 10:30. Sean Barrett is the narrator.

49. *Opfer der Vergangenheit*, 1937.

50. Burleigh, *Death and Deliverance*, 205–217.

51. Burleigh, *Death and Deliverance*, 207.

6. DARWINISM IN NAZI PROPAGANDA

1. Prof. Dr. B., "Darwin," *Rasse, Volk und Staat: Rassenhygienisches Beiblatt*, in *Völkischer Beobachter*, Norddeutsche Auflage (June 15, 1932).

2. Viktor Franz, "Das Göttliche im Gottesverneiner," *Völkischer Beobachter*, Norddeutsche Auflage Nr. 47 (February 16, 1934).

3. A. C., "Um die Abstammung des Menschen: Zum 20. Jahrestage Ernst Haeckels," *Völkischer Beobachter*, Norddeutsche Auflage (August 9, 1939): 6.

4. Heinz Brücher, "Ernst Haeckel, ein Wegbereiter biologischen Staatsdenkens," *Nationalsozialistische Monatshefte* 6 (1935): 1088–1096; see also Brücher's follow-up article, "Ernst Haeckel und die 'Welträtsel'-Psychose römischer Kirchenblätter," *Nationalsozialistische Monatshefte* 7 (1936): 261–265.

5. Heinz Brücher, "Rassen- und Artbildung durch Erbänderung, Auslese und Züchtung," *Nationalsozialistische Monatshefte* 12 (1941): 676. Emphasis in original.

6. Gerhard Heberer, "Abstammungslehre und moderne Biologie," *Nationalsozialistische Monatshefte* 7 (1936): 874–890.

7. Theodor Arzt, "Biologie und Weltanschauung," *Der Schulungsbrief* 6, no. 4 (1939): 149–155; quote at 155.

8. Ernst Lange, "Ludwig Woltmann," *Wille und Macht: Führerorgan der nationalsozialistischen Jugend* 6, no. 5 (1936): 8–11; see also review of Ludwig Woltmann, *Das Rassenwerk*, in *Wille und Macht* 6, no. 14 (1936): 62.

9. "Wertvolle Bücher," *Neues Volk* 3, 2 (1935): 46-47; "Bücher, die zu empfehlen sind," *Neues Volk* 3, no. 6 (1935): 46.

10. "Bücherecke," *Neues Volk* 5, no. 1 (1937): 47.

11. "Wissenschaft und Weltanschauung," *Neues Volk* 6, no. 12 (1938): 27.

12. Walter Gross, "Paläontologische Hypothesen zur Faktorenfrage der Deszendenzlehre: Über die Typen- und Phasenlehren von Schindewolf und Beurlen," *Die Naturwissenschaften* 31, no. 21/22 (1943): 237–245; quote at 237.

13. Isabel Heinemann, "Rasse, Siedlung, deutsches Blut," *Das Rasse- und Siedlungshauptamt der SS und die rassenpolitische Neuordnung Europas* (Göttingen: Wallstein Verlag, 2003), 162–163, 634–635.

14. Gerhard Heberer, "Neuere Funde zur Urgeschichte des Menschen und ihre Bedeutung für Rassenkunde und Weltanschauung," *Volk und Rasse* 12 (1937): 422–427, 435–444; quote at 444; see also Gerhard Heberer, "Jesuiten und Abstammungslehre," *Volk und Rasse* 13 (1938): 377–378; Heberer, "Die genetischen Grundlagen der Artbildung," *Volk und Rasse* 15 (1940): 136–137.

15. Eugen Fischer, "Die Entstehung der Menschenrassen," *Volk und Rasse* 13 (1938): 236.

16. Thomas Junker, *Die zweite Darwinische Revolution: Geschichte des Synthetischen Darwinismus in Deutschland* 1924 bis 1950 (Marburg: Basilisken-Presse, 2004), 256–257.

17. Christian von Krogh, "Schausammlung für Abstammungs- und Rassenkunde des Menschen in München," *Volk und Rasse* 13 (1938): 193–194.

18. Hans Weinert, "Zur Urgeschichte der nordischen und fälischen Rasse," *Rasse* 1 (1934): 345–355.

19. Hans Böker, "Rassenkonstanz—Artenwandel," *Rasse* 1 (1934): 250–254; Ludwig Plate, "Umweltlehre und Nationalsozialismus," *Rasse* 1 (1934): 279–283; Hans F. K. Günther, "Lebensgrundlage der Gattung Mensch und Verstädterung," *Rasse* 1 (1934): 49–56.

20. Kurt Holler, "Geologie und Umweltlehre," *Rasse* 2 (1935): 54–58.

21. Gisela Meyer-Heydenhagen, "Zum 80. Geburtstag des Grafen Georges Vacher de Lapouge," *Rasse* 2 (1935): 41–43; Hans F. K. Günther, "Zum Tode des Grafen Georges Vacher de Lapouge," *Rasse* 3 (1936): 95–98; Alexander Koch, "Ludwig Woltmann, der hervorragende Kämpfer für den Nordischen Gedanken," *Rasse* 3 (1936): 61–72.

22. Robert J. Richards, *Was Hitler a Darwinian?: Disputed Question in the History of Evolutionary Theory* (Chicago: University of Chicago Press, 2013), 239; see also Robert J. Richards, *The Tragic Sense of Life: Ernst Haeckel and the Struggle over Evolutionary Thought* (Chicago: University of Chicago Press, 2008), 446.

23. Kurt Hildebrandt, "Die Bedeutung der Abstammungslehre für die Weltanschauung," *Zeitschrift für die gesamte Naturwissenschaft* 3 (1937–38): 15–34; G. Hecht, "Biologie und Nationalsozialismus," *Zeitschrift für die gesamte Naturwissenschaft* 3 (1937–38): 280–290; quotes on 282, 285.

24. Richards, *Was Hitler a Darwinian?*, 239.

25. M. Westenhöfer, "Kritische Bemerkungen zu neueren Arbeiten über die Menschwerdung und Artbildung," *Zeitschrift für die gesamte Naturwissenschaft* 6, no. 3–4 (1940): 41–62.

26. Christian von Krogh, "Das 'Problem' der Menschwerdung," *Zeitschrift für die gesamte Naturwissenschaft* 6, no. 5–6 (1940): 105–112.

27. E. Bergdolt, "Abschliessende Bemerkungen zu dem Thema: 'Das Problem der Menschwerdung," *Zeitschrift für die gesamte Naturwissenschaft* 6, no. 7–8 (1940): 185–88.

28. *Rassenpolitik* (Berlin: Der Reichsführer SS, SS-Hauptamt, n.d. [approx. 1943]), 11–16, 21, 24–25, 27–28, 40, 50, 52, 64, 66.

29. *Rassenpolitik*, 61.

30. *Lehrplan für die weltanschauliche Erziehung in der SS und Polizei* (Berlin: SS-Hauptamt, n.d.), 78–79, 84.

31. *Wofür kämpfen wir?* (Berlin: Heerespersonalamt, 1944), iv–vi.

32. *Wofür kämpfen wir?*, 70.

33. *Wofür kämpfen wir?*, 110.

34. See Richard Weikart, *Hitler's Ethic*, 39–40, 43, 50–51, 91, 197; or Weikart, *Hitler's Religion*, passim, for other examples, mostly from Hitler.

7. Darwinist Ernst Haeckel in the Third Reich

1. For an excellent treatment of Haeckel's embryo drawings, see Jonathan Wells, *Icons of Evolution: Science or Myth?* (Washington, DC: Regnery, 2000), chap. 5.

2. Daniel Gasman, *The Scientific Origins of National Socialism: Social Darwinism in Ernst Haeckel and the German Monist League* (London: MacDonald, 1971); and Gasman, *Haeckel's Monism and the Birth of Fascist Ideology* (New York: Peter Lang, 1998); Robert J. Richards, *The Tragic Sense of Life: Ernst Haeckel and the Struggle over Evolutionary Thought* (Chicago: University of Chicago Press, 2008).

3. For more on Haeckel see Richard Weikart, *From Darwin to Hitler: Evolutionary Ethics, Eugenics, and Racism in Germany* (New York: Palgrave Macmillan, 2004).

4. Gasman, *Scientific Origins*, chap. 7, quote on 173.

5. Richards, *Tragic Sense of Life*, 445–446.

6. Robert J. Richards, *Was Hitler a Darwinian?: Disputed Question in the History of Evolutionary Theory* (Chicago: University of Chicago Press, 2013), 240.

7. Heinz Brücher, *Ernst Haeckels Bluts- und Geistes-Erbe: Eine kulturbiologische Monographie* (Munich: J. F. Lehmans Verlag, 1936), 91, 112–115.

8. Richard Weikart, "'Evolutionäre Aufklärung'? Zur Geschichte des Monistenbundes," in *Wissenschaft, Politik, und Öffentlichkeit: Von der Wiener Moderne bis zur Gegenwart*, eds. Mitchell G. Ash and Christian H. Stifter (Vienna: WUV Universitätsverlag, 2002), 131–148.

9. "Aufruf der Reichsarbeitsgemeinschaft freigeistiger Verbände," *Die Stimme der Vernunft: Monatshefte für wissenschaftliche Weltanschauung und Lebensgestaltung* 17 (1932): 54–56.

10. "Für die Freiheit des Gedankens," *Die Stimme der Vernunft: Monatshefte für wissenschaftliche Weltanschauung und Lebensgestaltung* 18 (1933): 23–24.

11. This point is made by Gereon Wolters, "Der 'Führer' und seine Denker: Zur Philosophie des 'Dritten Reichs,'" *Deutsche Zeitschrift für Philosophie* 47, no. 2 (1999): 223–251.

12. Hans Sluga, *Heidegger's Crisis: Philosophy and Politics in Nazi Germany* (Cambridge, MA: Harvard University Press, 1993), discusses in detail the role of idealism and existentialism in German philosophy during the Third Reich.

13. For Hitler's pantheistic religious views, see Richard Weikart, *Hitler's Religion: The Twisted Beliefs That Drove the Third Reich* (Washington, DC: Regnery History, 2016).

14. Adolf Hitler, monologue on October 14, 1941, in *Monologe im Führerhauptquartier 1941–1944: Die Aufzeichnungen Heinrich Heims*, ed. Werner Jochmann (Hamburg: Albrecht Knaus, 1980), 82–85; Hitler, "Adolf Hitlers Geheimrede vom 23. November 1937 auf der Ordensburg Sonthofen im Allgäu," in *Hitlers Tischgespräche im Führerhauptquartier*, ed. Henry Picker (Frankfurt: Ullstein, 1989), 487.

15. Erika Krausse and Uwe Hossfeld, "Das Ernst-Haeckel-Haus in Jena: Von der privaten Stiftung zum Universitätsinstitut," in *Repräsentationsformen in den biologische Wissenschaften*, eds. Armin Geus et al. (Berlin: VWB, 1999), 208–209.

16. Heinrich Schmidt, *Ernst Haeckel: Denkmal eines grossen Lebens* (Jena: Frommannsche Buchhandlung Walter Biedermann, 1934), v.

17. Krausse and Hossfeld, "Das Ernst-Haeckel-Haus in Jena," 203–231.

18. V. Franz, "Jenas Ernst-Haeckel-Haus und seine Aufgaben," *Der Biologe: Monatsschrift des Deutschen Biologen-Verbandes, des Sachgebietes Biologie des N.S.L.B.* 8 (1939): 261–264.

19. Karl Astel, "Geleitwort," in Heinz Brücher, *Ernst Haeckels Bluts- und Geistes-Erbe: Eine kulturbiologische Monographie* (Munich: J. F. Lehmans Verlag, 1936), 3–6.

20. Uwe Hossfeld, *Gerhard Heberer (1901-1973): Sein Beitrag zur Biologie im 20. Jahrhundert* (Berlin: Verlag für Wissenschaft und Bildung, 1997), 89.

21. Uwe Hossfeld, "Von der Rassenkunde, Rassenhygiene und biologischen Erbstatistik zur Synthetischen Theorie der Evolution: Eine Skizze der Biowissenschaften," in *Kämperische Wissenschaft: Studien zur Universität Jena im Nationalsozialismus*, eds. Uwe Hossfeld et al. (Cologne: Böhlau Verlag, 2003), 545–547.

22. Uwe Hossfeld, *Geschichte der biologischen Anthropologie in Deutschland: Von den Anfängen bis in die Nachkriegszeit* (Stuttgart: Franz Steiner Verlag, 2005), 252–257.

23. *Ausstellung grosse Deutsche in Bildnessen ihrer Zeit* (Berlin: Staatliche Museen und National-Galerie, 1936), 161. Haeckel's inclusion in this exhibition showed that the Nazis honored him, but we should not conclude from this that they agreed with him about everything, since the exhibit also included thinkers with mutually contradictory worldviews, e.g., Luther, Kant, Hegel, and Nietzsche.

24. Karl Richard Ganzer, ed., *Das deutsche Führergesicht: 204 Bildnisse deutscher Kämpfer und Wegsucher aus zwei Jahrtausenden* (Munich: J. F. Lehmanns Verlag, 1941), 192.

25. The list of approved texts is in "Biologielehrbücher für Höhere Schulen," in *Deutsche Wissenschaft Erziehung und Volksbildung: Amtsblatt des Reichsministeriums für Wissenschaft, Erziehung und Volksbildung und der Unterrichtsverwaltungen der Länder* 5 (1939): 460–461.

26. O. Steche, E. Stengel, and M. Wagner, *Lehrbuch der Biologie für Oberschulen und Gymnasien*, vol. 4 (for the 6., 7., and 8. Klasse) (Leipzig: Quelle und Meyer, 1942), 294 and plate after 296.

27. Jakob Graf, *Biologie für Oberschule und Gymnasium*, vol. 3: *Band für Klasse V, Der Mensch und die Lebensgesetze* (Munich: J. F. Lehmanns Verlag, 1940), 9.

28. Karl Kraepelin, C. Schaeffer, and G. Franke, *Das Leben*, vol. 4B: *Klasse 6 bis 8 der Mädchenschulen* (Leipzig: B. G. Teubner, 1942), 255.

29. Hans-Christian Harten, Uwe Neirich, and Matthias Schwerendt, *Rassenhygiene als Erziehungsideologie des Dritten Reichs: Bio-bibliographisches Handbuch* (Berlin: Akademie Verlag, 2006), 233; also see the cover and list of editors of any copy of *Der Biologe* 10 (1941).

30. Ferdinand Rossner, *Der Weg zum ewigen Leben der Natur: Gegenwartsfragen der biologischen Lebenskunde*, 2nd ed. (Langensalza: Verlag von Julius Beltz, 1937), 9–10. The Nazi Education Ministry approved this book in "Verzeichnis der zur Beschaffung für Schulbüchereien (Lehrer- und Schülerbüchereien) geeigneten Bücher und Schriften," in *Deutsche Wissenschaft Erziehung und Volksbildung: Amtsblatt des Reichsministeriums für Wissenschaft, Erziehung und Volksbildung und der Unterrichtsverwaltungen der Länder* 4 (1938): 396.

31. G. Heberer, "Haeckel, Ernst," in *Handbuch für den Biologieunterricht*, ed. Ferdinand Rossner (Langensalza: Julius Beltz, n.d.), 208–209.

32. Ferdinand Rossner, *Was Wir vom Leben wissen: Grundfragen der Biologie* (Braunschweig: Georg Westermann, 1942), 41, 77.

33. Werner Haeckel (great nephew of Ernst), "Ernst Haeckel und die Gegenwart," *Der Biologe: Monatsschrift zur Wahrung der Belange der Biologie und der deutschen Biologen* 3 (1934): 33–34.

34. Friedrich Lipsius, "Ernst Haeckel als Naturphilosoph," *Der Biologe: Monatsschrift zur Wahrung der Belange der Biologie und der deutschen Biologen* 3 (1934): 43–46.

35. Gerhard Heberer, "Der gegenwärtigen Vorstellungen über den Stammbaum der Tiere und die "Systematische Phylogenie" E. Haeckels," *Der Biologe* 8 (1939): 264–273; Werner Zündorf, "Ernst Haeckels Stammbaum der Pflanzen. Zum 20. Todestag Ernst Haeckels,"

Der Biologe: Monatsschrift des Deutschen Biologen-Verbandes, des Sachgebietes Biologie des N.S.L.B. 8 (1939): 273–279.

36. F. Rossner, "Systematik und Entwicklungsgedanke im Unterricht," *Der Biologe: Monatsschrift des Deutschen Biologen-Verbandes, des Sachgebietes Biologie des N.S.L.B.* 8 (1939): 366–372.

37. Viktor Franz, "Die Fortschritts- oder Vervollkommnungstheorie, der Aufbau auf Haeckels Stammesgeschichte," *Archiv für Rassen- und Gesellschafts-Biologie* 31 (1937): 281–282.

38. Richards, *Tragic Sense of Life*, 446.

39. G. Hecht, "Biologie und Nationalsozialismus," *Zeitschrift für die gesamte Naturwissenschaft* 3 (1937): 280–290; quotes at 282, 285. Emphasis in original.

40. Kurt Hildebrandt, "Die Bedeutung der Abstammungslehre für die Weltanschauung," *Zeitschrift für die gesamte Naturwissenschaft* 3 (1937): 15–34.

41. Katja Geisenhainer, *"Rasse ist Schicksal": Otto Reche (1879–1966) – ein Leben als Anthropologe und Völkerkundler* (Leipzig: Evangelische Verlagsanstalt, 2002), 180–181.

42. Hossfeld, *Geschichte der biologischen Anthropologie in Deutschland*, 326.

43. Quoted in Ute Deichmann, *Biologists under Hitler*, trans. Thomas Dunlap (Cambridge, MA: Harvard University Press, 1996), 270; ellipses in Deichmann.

44. Walter Gross, "Paläontologische Hypothesen zur Faktorenfrage der Deszendenzlehre: Über die Typen- und Phasenlehren von Schindewolf und Beurlen," *Die Naturwissenschaften* 31, no. 21/22 (1943): 237–245.

45. Gerhard Heberer, *Ernst Haeckel und wissenschaftliche Bedeutung: Zum Gedächtnis der 100. Wiederkehr seines Geburtstages* (Tübingen: Akademische Verlagsbuchhandlung Franz F. Heine, 1934), iv, 14, 28.

46. Hossfeld, *Gerhard Heberer*, 52.

47. Gerhard Heberer, "Neuere Funde zur Urgeschichte des Menschen und ihre Bedeutung für Rassenkunde und Weltanschauung," *Volk und Rasse* 12 (1937): 422–427; Heberer, "Jesuiten und Abstammungslehre," *Volk und Rasse* 13 (1938): 377–378.

48. Clipping from *Deutsche Zeitung in den Niederlanden*, April 8, 1942, Hoover Archives, Stanford University, Germany Deutsche Kongress-Zentrale Records, 1870–1943, Box 136 – Haeckel-Gesellschaft, 1942; see also Hossfeld, *Geschichte der biologischen Anthropologie in Deutschland*, 256, n. 165.

49. Hans Weinert, *Ursprung der Menschheit: Über den engeren Anschluss des Menschengeschlechts an die Menschenaffen*, 2nd ed. (Stuttgart: Fredinand Enke Verlag, 1944), v, 4–5, 314.

50. Katja Geisenhainer, *"Rasse ist Schicksal": Otto Reche (1879–1966) – ein Leben als Anthropologe und Völkerkundler* (Leipzig: Evangelische Verlagsanstalt, 2002), 27, 46.

51. Hossfeld, *Geschichte der biologischen Anthropologie in Deutschland*, 255.

52. Hans F. K. Günther, "Die Erneuerung des Familiengedankens in Deutschland," in *Führeradel durch Sippenpflege: Vier Vorträge* (Munich: J. F. Lehmanns Verlag, 1936), 46. This book by Günther was dedicated to the Nazi Minister of the Interior Wilhelm Frick and the copy I read was owned originally by the Adolf-Hitler-Schule Bibliothek.

53. Günther, "Die Notwendigkeit einer Führerschicht für den völkischen Staat," in *Führeradel durch Sippenpflege*, 91.

54. Otto Ammon, *Die natürliche Auslese beim Menschen* (Jena: Gustav Fischer, 1893); *Die Gesellschaftsordnung und ihre natürliche Grundlagen*, 3rd ed. (Jena: Gustav Fischer, 1900); Alexander Tille, *Volksdienst* (Berlin: Wiener'sche Verlagsbuchhandlung, 1893); *Von Darwin bis Nietzsche: Ein Buch Entwicklungsethik* (Leipzig: C. G. Naumann, 1895).

55. Günther, "Vererbung und Erziehung," in *Führeradel durch Sippenpflege*, 106.

56. V. Franz, "Das Göttliche im Gottesverneiner," *Völkischer Beobachter*, Norddeutsche Auflage Nr. 47 (February 16, 1934).

57. A. C., "Um die Abstammung des Menschen: Zum 20. Jahrestage Ernst Haeckels," *Völkischer Beobachter*, Norddeutsche Auflage (August 9, 1939): 6.

58. Heinz Brücher, "Ernst Haeckel, ein Wegbereiter biologischen Staatsdenkens," *Nationalsozialistische Monatshefte* 6 (1935): 1088–1096; quote at 1092.

59. Heinz Brücher, "Ernst Haeckel und die 'Welträtsel'-Psychose römischer Kirchenblätter," *Nationalsozialistische Monatshefte* 7 (1936): 261–265.

60. Ernst Lange, "Ludwig Woltmann," *Wille und Macht: Führerorgan der nationalsozialistischen Jugend* 6, no. 5 (1936): 8–11; quote on 9.

61. "Richtlinien für die Bestandsprüfung in den Volkbüchereien Sachsens," *Die Bücherei: Zeitschrift fürt deutsche Schrifttumspflege* 2 (1935): 279–280.

62. Richards, *Was Hitler a Darwinian?*, 240.

63. For Nazi lists of banned books that do not include Haeckel's name, see http://www.berlin.de/rubrik/hauptstadt/verbannte_buecher/index.php; accessed September 9, 2015.

64. *Ernst Haeckel: Sein Leben, Denken und Wirken: Eine Schriftenfolge für seine zahlreichen Freunde und Anhänger*, vol. 1, ed. Victor Franz (Jena: Verlag von Wilhelm Gronau, W. Agricola, 1943), 160.

65. Ernst Krieck, *Leben als Prinzip der Weltanschauung und Problem der Wissenschaft* (Leipzig: Armanen-Verlag, 1938), 8, 24, 26, 88, 158–159, 167–168.

66. Krausse and Hossfeld, "Das Ernst-Haeckel-Haus in Jena," 214; Deichmann, *Biologists under Hitler*, 270.

67. Sluga, *Heidegger's Crisis*, 223–24.

68. On Haeckel and anti-Semitism, see Stefan Wogawa, Uwe Hossfeld, and Olaf Breidbach, "'Sie ist eine Rassenfrage': Ernst Haeckel und der Antisemitismus," in *Anthropologie nach Haeckel*, ed. Dirk Preuss, Uwe Hossfeld, and Olaf Breidbach (Stuttgart: Franz Steiner Verlag, 2006), 220–240.

8. DARWINISM IN NEO-NAZISM AND WHITE NATIONALISM

1. Anton Lavey, foreword to *Might Is Right*, by Ragnar Redbeard (Amazon Digital Services, 2012), Kindle.

2. Andrew Hilton, "Book Review: Might Is Right or the Survival of the Fittest, by Ragnar Redbeard," Occidental Observer, September 29, 2009, www.theoccidentalobserver.net/2009/09/29/book-review-might-is-right-or-the-survival-of-the-fittest-by-ragnar-redbeard/.

3. James Harting, "Ragnar Redbeard, Friedrich Nietzsche, and the Will-to-Power," Storm Front Forum, September 20, 2019, updated on October 17, 2019, www.stormfront.org/forum/blogs/u224711-e4764/.

4. Ragnar Redbeard, *Might Is Right or The Survival of the Fittest* [1896] (Port Townsend, WA: Loompanics Unlimited, c. 1974), 5, 8–9, 35, 17–18, 47–48.

5. Redbeard, *Might Is Right*, 18, 77.

6. Redbeard, *Might Is Right*, 77.

7. Redbeard, *Might Is Right*, 32, 91.

8. George Lincoln Rockwell, *White Power*, "Chapter 15: National Socialism," accessed February 6, 2020, www.americannaziparty.com/white_power.pdf.

9. Rockwell, *White Power*, "Chapter 3: The Chart Forgers."

10. Rockwell, *White Power*, "Chapter 9: The Black Plague."

11. Rockwell, *White Power*, "Chapter 10: The Facts of Race."

12. Rockwell, *White Power*, "Chapter 10: The Facts of Race."

13. "What Is National Socialism?," New Order, accessed February 19, 2020, www.theneworder.org/National-Socialism.html.

14. "Introduction to the New Order," New Order, accessed February 19, 2020, www.theneworder.org/Intro%20New%20Order0001.pdf.

15. Damon T. Berry, *Blood and Faith: Christianity in American White Nationalism* (Syracuse, NY: Syracuse University Press, 2017), 44.

16. William Pierce, "Cosmotheism, Wave of the Future," *National Vanguard*, September 11, 2019 (but based on a speech from 1977), nationalvanguard.org/2019/09/wlp86-william-pierce-on-cosmotheism-wave-of-the-future/.

17. "What Is the National Alliance?," National Alliance, accessed February 18, 2020, natall.com/about/what-is-the-national-alliance/.

18. For more on Ratzel, see Richard Weikart, *From Darwin to Hitler: Evolutionary Ethics, Eugenics, and Racism in Germany* (New York: Palgrave Macmillan, 2004), 112–114, 192–194.

19. "About Us," *National Vanguard*, accessed February 18, 2020, nationalvanguard.org/about/. Emphasis in original.

20. "About Us," *National Vanguard*.

21. Kevin Alfred Strom, "Our Evolutionary Morality," *National Vanguard*, July 14, 2018, nationalvanguard.org/2018/07/our-evolutionary-morality/.

22. Strom, "Our Evolutionary Morality."

23. Richard McCulloch, "Richard McCulloch's The Racial Compact—Part One—The Evolutionary Basis of Racial Diversity," *National Vanguard*, September 24, 2019, nationalvanguard.org/2019/09/richard-mccullochs-the-racial-compact-part-one-the-evolutionary-basis-of-racial-diversity/.

24. James Hart, "God, Evolutionary Ethics, and Eugenics," *National Vanguard*, October 31, 2015, nationalvanguard.org/2015/10/god-evolutionary-ethics-and-eugenics/.

25. H. Millard, "They Want to Stop Our Evolution," *National Vanguard*, accessed February 11, 2020, nationalvanguard.org/2019/02/they-want-to-stop-our-evolution/.

26. H. Millard, "Whites Should Reject the False Idea That Hitler and National Socialism Were 'Evil,'" *National Vanguard*, June 28, 2017, nationalvanguard.org/2017/06/whites-should-reject-the-false-idea-that-hitler-and-national-socialism-were-evil/.

27. Rosemary Pennington, "Human Origins: Multiregional or 'Out of Africa'?," *National Vanguard*, accessed February 11, 2020, nationalvanguard.org/2019/04/human-origins-multiregional-or-out-of-africa/.

28. Human Evolution News, accessed February 24, 2020, www.subspecieist.com.

29. Berry, *Blood and Faith: Christianity in American White Nationalism*, 19.

30. Revilo P. Oliver, "To Honor Darwin," *National Vanguard*, April 9, 2017, nationalvanguard.org/2017/04/to-honor-darwin/.

31. Revilo P. Oliver, "The Piltdown Forgery," *National Vanguard*, October 29, 2017, nationalvanguard.org/2017/10/the-piltdown-forgery/.

32. "What Is Cosmotheism?," Cosmotheism: Toward a New Consciousness, accessed February 19, 2020, cosmotheistchurch.org.

33. Berry, *Blood and Faith*, 77.

34. Ben Klassen, *Nature's Eternal Religion* (Milwaukee, WI: Milwaukee Church of the Creator, 1992), 5, 15–16, 189, accessed February 6, 2020, www.daemuk.ch/ebooks/nature-s-eternal-religion-ben-klassen-english-pdf-ebook.pdf.

35. Klassen, *Nature's Eternal Religion*, 6, 15–16, 194.

36. Jared Taylor, "A New Theory of Racial Differences," *American Renaissance* 5, no. 12, December 1994, www.amren.com/archives/back-issues/december-1994/.

37. "An American Renaissance Reading List," American Renaissance, January 15, 2017, www.amren.com/commentary/2017/01/alt-right-reading-list-jared-taylor-white-nationalist-books/.

38. Hubert Collins, "The Decline of Western Man (and Woman)," American Renaissance, October 31, 2014, https://www.amren.com/features/2014/10/the-decline-of-western-man-and-woman/.

39. There is much literature refuting Rushton's racist psychology. See, for instance, Christopher Jencks and Meredith Phillips, "The Black-White Test Score Gap: Why It Persists and What Can Be Done," Brookings, March 1, 1998, https://www.brookings.edu/articles/the-black-white-test-score-gap-why-it-persists-and-what-can-be-done/; and Robert Sternberg, Elena Grigorenko, and Kenneth Kidd, "Intelligence, Race, and Genetics," *American Psychologist* 60 (2005): 46–59.

40. J. Philippe Rushton, *Race, Evolution, and Behavior: A Life History Perspective*, 2nd special abridged edition (Port Huron, MI: Charles Darwin Research Institute, 2000), 7; accessed on Identity Evropa website (which is no longer functional), October 14, 2019.

41. Rushton, *Race, Evolution, and Behavior*, 7–12, passim.

42. Collins, "The Decline of Western Man (and Woman)."

43. Andrew Anglin, "A Normie's Guide to the Alt-Right," The Daily Stormer, August 31, 2016, www.dailystormer.com/a-normies-guide-to-the-alt-right.

44. Kevin MacDonald, *Separation and Its Discontents: Toward an Evolutionary Theory of Anti-Semitism* (Westport, CT: Praeger, 1998), viii.

45. Kevin MacDonald, *The Culture of Critique: An Evolutionary Analysis of Jewish Involvement in Twentieth-Century Intellectual and Political Movements* (Westport, CT: Praeger, 1998), vii.

46. MacDonald, *Culture of Critique*, 17.

47. MacDonald, *Culture of Critique*, 323.

48. Kevin MacDonald, "Pat Buchanan on Darwin," Occidental Observer, accessed February 7, 2020, www.theoccidentalobserver.net/2009/07/01/pat-buchanan-on-darwin/.

49. Kevin MacDonald, "Ben Stein's *Expelled*: Was Darwinism a Necessary Condition of the Holocaust?," Occidental Observer, accessed February 7, 2020, www.theoccidentalobserver.net/2008/12/01/ben-steins-expelled-was-darwinism-a-necessary-condition-for-the-holocaust/.

50. Thomas J. Main, *The Rise of the Alt-Right* (Washington, DC: Brookings Institution Press, 2018), chap. 1.

51. Richard Spencer and F. Roger Devlin, "Race—Stalking the Wild Taboo," National Policy Institute website, October 5, 2017, https://nationalpolicy.institute/2017/10/05/race-stalking-the-wild-taboo/.

52. Guillaume Durocher, "An Uncertain Idea of Europe," Radix Journal, June 30, 2016, https://www.radixjournal.com/2016/06/2016-6-30-an -uncertain-idea-of-europe.

53. Guillaume Durocher, "Biocentric Political Thought in the Third Reich: A Review of Johann Chapoutot's The Law of Blood," Occidental Observer, accessed February 4, 2020, https://www.theoccidentalobserver.net/2018/12/15/biocentric-political-thought-in-the-third-reich-a-review-of-john-chapoutots-the-law-of-blood/.

54. "Madison Grant and the American Nation," Radix Journal, October 8, 2016, radixjournal.com/2016/10/2016-10-6-madison-grant-and-the-american-nation/.

55. Jared Taylor, "Killing the Messenger," *American Renaissance*, June 1992, www.amren.com/archives/back-issues/june-1992/#article2.

56. Jared Taylor, "Race Differences in Intelligence," American Renaissance, October 27, 2019, www.amren.com/news/2019/10/race-differences-in-intelligence-richard-lynn/.

57. Jared Taylor, "The Human Animal," *American Renaissance* 20, no. 1, January 2009, 7–13, www.amren.com/ar/pdfs/2009/200901ar.pdf.

58. For 2021 rape statistics by country, see World Population Review, https://worldpopulationreview.com/country-rankings/rape-statistics-by-country.

59. Guillaume Durocher, "Darwin on the Rise and Fall of the Human Races," Occidental Observer, January 12, 2019, www.theoccidentalobserver.net/2019/01/12/darwin-on-the-rise-and-fall-of-human-races-part-1-of-2/; Michael Levin, "The Evolution of Racial Differences in Morality," *American Renaissance* 6, no. 4, April 1995, www.amren.com/archives/back-issues/april-1995/; F. Roger Devlin, "Ed Dutton with an Evolutionary Perspective on the Rape of Finland," Occidental Observer, March 20, 2019, www.theoccidentalobserver.net/2019/03/20/ed-dutton-with-an-evolutionary-perspective-on-the-rape-of-finland/.

60. Frank Hilliard, "The Alternative Right Belongs to the Darwinians," Council of European Canadians, April 24, 2016, at https://www.eurocanadian.ca/2016/04/alternative-right-belongs-to-darwinians.html.

9. Conclusion

1. Otto Reche, "Die Genetik der Rassenbildung beim Menschen," in *Die Evolution der Organismen: Ergebnisse und Probleme der Abstammungslehre*, ed. Gerhard Heberer (Jena: Gustav Fischer, 1943), 705.

2. Benoit Massin, "Rasse und Vererbung als Beruf: Die Hauptforschungsrichtungen am Kaiser-Wilhelm-Institut für Anthropologie, menschliche Erblehre und Eugenik im Nationalsozialismus," in *Rassenforschung an Kaiser-Wilhelm-Instituten vor und nach 1933*, ed. Hans-Walter Schmuhl (Göttingen: Wallstein Verlag, 2003), 194.

3. Sir Arthur Keith, *Evolution and Ethics* (New York: G. P. Putnam's Sons, 1946), 9–10.

4. Edward Dutton, "Jewish Group Evolutionary Strategy Is the Most Plausible Hypothesis," *Evolutionary Psychological Science* 5 (2019): 136–142.

Figure Credits

Figure 1.1. Frontispiece of Ernst Haeckel's 1868 book showing humans and apes. Haeckel, *Natürliche Schöpfungsgeschichte* (Berlin: Georg Reimer, 1868).

Figure 2.1. Hitler's message in a book on Christmas: "All of nature is a powerful struggle between power and weakness, an eternal victory of the strong over the weak." *Deutsche Kriegsweihnacht*, 3rd ed. (Munich: Zentralverlag der NSDAP, Franz Eher Verlag, 1943).

Figure 3.1. Sepp Burgstaller's picture book. Burgstaller, *Hereditary Theory, Racial Science, and Population Policy: 400 Picture Sketches for Use in Schools [Erblehre, Rassenkunde und Bevölkerungspolitik]* (Vienna: Deutscher Verlag für Jugend und Volk, 1941), 29.

Figure 4.1. Phylogenetic tree of human races in anthropologist Hans Weinert's *Biologische Grundlagen für Rassenkunde und Rassenhygiene* (Stuttgart: Ferdinand Enke, 1934).

Figure 4.2. Evolutionary tree of primates that includes humans in anthropologist Hans Weinert's *Entstehung der Menschenrassen* (Stuttgart: Fredinand Enke Verlag, 1941).

Figure 5.1. Nazi school poster: "Elimination of the Sick and Weak in Nature." Alfred Vogel, *Erblehre und Rassenkunde in bildlicher Darstellung* (Stuttgart: Verlag für Nationale Literatur Gebr. Rath, 1938).

Figure 6.1. Nazi periodical cover with the message "Life Requires Struggle." *Der Schulungsbrief* 9, no. 4 (1942).

INDEX